AF888325

SOTHIS PRESS

Sothis Press aims to completely reimagine the public domain book with an emphasis on universal access to captivatingly designed print editions. We believe that technology will enable every single book ever published to be perpetually accessible in beautiful print form. Sothis is beauty in ubiquity.

www.sothis.press
1 Waterhouse Square, London, ECIN 2ST

Motion Study

A Method for Increasing the
Efficiency of the Workman

Frank Bunker Gilbreth

SOTHIS PRESS

ISBN 978-1-0365-1742-7

Motion Study is in the public domain.

This edition is published by Sothis Press The company is registered in England and Wales. The Company Number is 14900389.

MOTION STUDY

A METHOD FOR INCREASING THE EFFICIENCY OF THE WORKMAN

BY

FRANK B. GILBRETH

MEMBER OF THE AMERICAN SOCIETY OF MECHANICAL ENGINEERS

WITH AN INTRODUCTION BY

ROBERT THURSTON KENT

EDITOR OF "INDUSTRIAL ENGINEERING"

SIXTH THOUSAND

NEW YORK:
D. VAN NOSTRAND COMPANY
EIGHT WARREN STREET
1921

PREFACE

THE phrase "Motion Study" explains itself.

The aim of motion study is to find and perpetuate the scheme of perfection. There are three stages in this study:

1. Discovering and classifying the best practice.

2. Deducing the laws.

3. Applying the laws to standardize practice, either for the purpose of increasing output or decreasing hours of labor, or both.

Standardizing the trades is the world's most important work to-day, and motion study is the first factor in that work.

In presenting this material I have attempted to show the necessity for Motion Study and the savings that are possible by the application of its underlying principles.

Thanks are due to the Myron C. Clark Publishing Company and to *Industrial Engineering* for permission to use the cuts that illustrate this book.

FRANK B. GILBRETH.

TABLE OF CONTENTS

LIST OF ILLUSTRATIONS

INTRODUCTION

WHEN the editor of a live journal hears of some new development in the field to which his paper caters, he is neglecting his duty if he does not make every possible effort to secure a powerful article by the highest authority on that development for his readers. Some months before "Motion Study" first appeared in the columns of *Industrial Engineering*, we heard that Mr. Gilbreth had, by some method or other, made wonderful records in the construction of buildings and other engineering works. We were curious to know how this had been accomplished, and sought an interview. We then learned that for years Mr. Gilbreth had been studying the actions of his workmen, the conditions surrounding their work, and all the other variables which go to help or hinder them in the actual construction work. He had paid particular attention to the motions made by a given man, say a bricklayer, in getting a brick from the pile made by the tenders to its final resting place in the wall. He had discovered how the number of these motions could be cut down, by having the brick brought to the bricklayer in a different way than usual, by placing them in a somewhat more convenient position, by arranging the scaffolding in an improved manner, and by various other little changes, some of them slight in themselves, but all together totaling an immense saving of time and motions.

This appealed to us as forming the basis of a good "story," and we immediately made arrangements with Mr. Gilbreth to prepare for *Industrial Engineering* a serial article showing the importance of these studies in the improvement of working conditions in all trades, and how they benefited both employer and workman. When the manuscript of "Motion Study" was first placed in the writer's hands, as editor, he examined it with considerable interest, but without any idea of the immense importance of what he had done in securing this contribution. It was then merely another good story for the paper. It seemed to him a trifle unfortunate, however, that the author had chosen practically all of his illustrations from the bricklaying trade. On rereading the article more carefully it was apparent that these illustrations were of secondary importance and had been drawn from a trade with which every one was familiar and which any one could observe, also from one so old that it hardly seemed possible that it could be improved. It therefore became evident that the *principles* laid down were applicable to every trade and industry. This idea was confirmed later, when "Motion Study" appeared in serial form.

We were a trifle disappointed, at first, that it attracted less attention than we anticipated. We thought that perhaps we were somewhat in advance of our time, and that the public was not yet prepared for so long a step forward. It now appears, however, that the apparent lack of interest was due to the fact that we had presented a subject so entirely new that it required some little time for people to comprehend its importance and to realize its value. By the time the third installment had appeared, requests for the earlier installments were flowing in steadily, and since its completion many firms have sent for all the issues of the paper containing it. To show the wide application possible of the principles laid down in "Motion Study," requests for the complete series have come to us from the iron and steel industry, from the shoe manufacturing industry, from book-printing and book-binding establishments, and many other industries. It was when we began to receive these requests that we realized that we had done something worth while, and had published an article which was of stupendous value, not to one trade or group of trades, but to the whole world.

The writer, in handling the successive installments of "Motion Study," became more and more impressed with the possibilities which were involved in it. He resolved to apply some of these principles in his own office. Naturally the first point of attack is the one where the greatest saving can be accomplished. In our case, it happened to be the outgoing mail. A publication has, particularly in its circulation department, an amount of outgoing mail entirely out of proportion to the volume of business transacted by it, when measured by the standard of other industries. A circulation campaign will involve the sending out of perhaps twenty thousand duplicate letters, each with one or more inclosures. Evidently the saving of but one motion on each letter would, in the aggregate, show an immense saving of time. Formerly the girls folding and sealing the letters were permitted to arrange the work to suit themselves. A short observation of their work showed that there was much room for improvement. The writer studied the question for a short time and made several experiments to determine in just what order each movement should be made to fold the letter, pick up its inclosure, pick up the envelope, and insert the letter and inclosure in the envelope. The first attempt was crude, but immediately doubled the output of the girl. Further study resulted in improvements which not only eliminated motions, but shortened the distance which the hands had to move in those that remained. The final result was an arrangement of pieces and a sequence of motions by which each hand, at the completion of one motion, was in position immediately to begin the next. The final motion, that of throwing the filled envelope on the pile, was eliminated entirely by having a large basket on the floor, directly under the point where the letter was inserted in the envelope. The girl simply let go of the envelope, and it fell into the basket, gravity doing the work formerly done by the girl. The output under the new conditions was about four times that obtained when the girls were allowed to do the work their own way.

Several other routine jobs in the office were handled in

the same way, with a marked improvement in each case. By this time the office force had become interested and were studying motions on their own account, and improving methods of doing work without any suggestion from the writer. One of the girls devised a method of stamping envelopes which enabled her to work at a speed of between one hundred and one hundred and twenty envelopes per minute. She piled the letters on edge in a long pile, the addressed side facing her. The stamps were torn in strips, crosswise of the sheet, so that the stamps were side by side instead of one above the other. She fastened to the forefinger of her right hand a small wet sponge, and taking a strip of stamps in that hand, fed them across the sponge, using her thumb to move the strip, and to guide the stamp into place on the corner of the envelope. The left hand drew the stamped envelope forward from the pile, the thumb of the left hand giving the necessary pressure to the stamp while it was being drawn forward, to assure its being firmly affixed to the envelope. The motion of drawing the envelope forward tore the stamp from the strip, and the operation was complete. The work was done with marvelous rapidity, yet the girl hardly seemed to make any motions, except to pull the envelopes forward, and to reach for strips of stamps. We do not know just what processes were followed in developing the method, as the girl studied it out and put it in operation while the writer was taking a vacation.

These incidents, and others, convinced us that there was

much more to motion study than appeared from a hasty survey of the subject. We then began to look around in earnest, to discover what had been done in this line in other trades than those with which Mr. Gilbreth was familiar. We found that practically nothing had been done in a systematic, scientific manner, except in certain shops where scientific management had been installed.

We further found, that even in these shops motion study had not been made in the scientific manner outlined in Mr. Gilbreth's articles. It was a by-product, an incident in the installation of scientific management, rather than a science of itself. Nevertheless, even treated as an incidental branch of management it had conferred much benefit on those shops in which it had been made. We shall refer to this later.

The reduction of the number of motions can be accomplished in two radically different ways: (1) By analyzing every step of a process, as outlined by Mr. Gilbreth, studying the motions made, and improving or eliminating them as a result of the analysis, or by devising an entirely new way of accomplishing the same object. (2) By substituting a device which is an improvement over that formerly used, but which required a greater number of motions to operate it, or by the substitution of new motions or processes as they occur to the observer, rather than by any systematic study of the subject. This last method is the one most generally used. It might be termed the "accidental" method, as contrasted with the scientific one developed by Mr. Gilbreth.

An example of this latter kind of motion study is familiar to every man, woman, or child who lives in even a moderate-sized city. In the city fire departments much thought has been expended on the problem of enabling the firemen to start out to a fire in the shortest possible time after the alarm has been received. At first the horses stood in their stalls with their harness on them. The hitching of the horses required the fastening of several buckles. Some one then invented the drop harness, now universally used, and the number of motions in hitching a horse was reduced to three, — snapping the collar round his neck, and the fastening of the two reins to his bridle by bit snaps. Later the horses were moved from the rear of the house to a point alongside the engine, so that they had only to travel a matter of a few feet to be in position under the harness. Some one else then invented a device which released the horses from their stalls automatically with the sounding of the alarm on the fire-house gong, effecting a further saving in time and motion. Instead of having the firemen descend stairs from their sleeping quarters, the sliding pole was thought of, which eliminated a great number of individual motions and saved many seconds. And so on, as one device after another was perfected which saved motions, and thereby time, it was adopted, until now the "motion efficiency" in a fire house is one hundred per cent. It is useless to improve it further, because it has reached a point where the company is ready to start to a fire before it has received the number of the box. The men and horses do their part in less time than the electric telegraph transmits the complete signal. Yet to attain this efficiency has required a period of perhaps thirty years. The subject was not studied in a scientific manner.

Turning now to the machine shop, let us see what motion study means there. In an editorial in *Industrial Engineering*, in August, 1910, we said:

Before a task can be set intelligently it is necessary to know just what can be accomplished by the best type of workman. This usually involves a time study of the job under consideration. The time study is more than putting a clerk with a stop watch alongside the workman, with instructions to see how long it takes him to do the job. A proper time study requires that a certain piece of work be divided into its component operations and that each operation be studied separately, and also in conjunction with other operations to which it is related. The time of performing these operations is recorded not once, but many times, until a fair average has been determined. The results are then analyzed to see if the time required can be cut down. Usually it will be found that it can. A single instance will suffice for illustration.

In a certain shop with which we are familiar a piece had to have several holes of different sizes drilled in it, a jig being provided to locate the holes. The drills and the sockets for them were given to the workman in a tote box. The time study of this job revealed several interesting facts. First, after the piece was drilled the machine was stopped, and time was lost while the workman removed the piece from the jig and substituted a new one. This was remedied by providing a second jig in which the piece was placed while another piece was being drilled in the first jig, the finished one being removed after the second jig had been placed in the machine and drilling started. It was also found that the workman lost considerable time hunting in the tote box for his drill, and for the socket to fit it. The result was the provision of a socket for each drill, which was fitted to it in the tool room, and the further provision of a tray alongside the machine on which the man could lay out his drills in the order in which they were to be used. He was thereby enabled to pick up the correct drill without losing any time hunting for it. It was also found that it took considerable time to verify the size of the drill and socket, due to the figures stamped on the drill by the manufacturer being so small as to require the workman to go to the light to hunt for the figures. Consequently, numbers one-half inch high were placed on the drills, so that they could be seen in almost any light. To do this it was necessary to grind a flat spot on the drill to accommodate the large-size figures. This had the desirable but unsought for result of enabling the workman to locate the figures by the sense of touch, and consequently no time was lost in turning the drill round and round to search for the size.

The above changes, simple in themselves, resulted in an increase in output on this particular job of about four times that usually obtained before the time study was made.

The incident related above is one that the writer discovered after he began to investigate motion study in other lines than those discussed by Mr. Gilbreth. It, too, is typical of the "accidental" method. It is sufficient to show, however, what scientific motion study could do if applied to an entire industry.

The following letter from Mr. H. L. Gantt to the writer illustrates how little people, even the most expert in their line, know about the most economical way of doing work:

Editor Industrial Engineering:

The series of articles on "Motion Study" by Frank B. Gilbreth are particularly valuable as illustrating what a man, having an analytical mind and a quick comprehension of details, can accomplish. They also illustrate the fact that in order to accomplish such results the subject must be given a great deal of study. The articles are unique in that they describe for the first time the detailed application to an art of principles which have often been described in more general terms.

On the other hand, the man who becomes interested in making this kind of a study will, if he has the right kind of mind, become so fascinated by it that it is on his mind almost all the time. The subject opens up so many possibilities that those men who can appreciate it are simply carried away. While in London with the American

Society of Mechanical Engineers, Mr. Gilbreth cornered an old friend of his and explained to him the wonderful results that could be accomplished by motion study. He declared that he did not care what the work was, he would be able to shorten the time usually required, provided that nobody had previously applied the principles of motion study to the work.

A few days before, this friend had been at the Japanese-British Exposition and had seen there a girl putting papers on boxes of shoe polish at a wonderful speed. Without saying what he had in mind, Mr. Gilbreth's friend invited him to visit the exposition, and in a most casual way led him to the stand where the girl was doing this remarkable work, with the feeling that here at least was an operation which could not be improved upon.

No sooner had Mr. Gilbreth spied this phenomenal work than out came his stop watch, and he timed accurately how long it took the girl to do twenty-four boxes. The time was forty seconds. When he had obtained this information he told the girl that she was not doing the work right. She, of course, was greatly incensed that a man from the audience should presume to criticize what she was doing, when she was acknowledged to be the most skillful girl that had ever done that work. He had observed that while all her motions were made with great rapidity, about half of them would be unnecessary if she arranged her work a little differently. He has a very persuasive way, and although the girl was quite irritated by his remark, she consented to listen to his suggestion that he could show her how to do the work more rapidly. Inasmuch as she was on piece work the prospect of larger earnings induced her to try his suggestion. The first time she tried to do as he directed she did twenty-four boxes in twenty-six seconds; the second time she tried it she did it in twenty seconds. She was not working any harder, only making fewer motions.

This account the writer heard in Manchester, England, from the man himself who had put up the job on Mr. Gilbreth, and it is safe to say that this man is now about as firm a believer in motion study as Mr. Gilbreth is.

H. L. GANTT.

NEW YORK, Oct. 1, 1910.

Enough has been said, and sufficient instances drawn from widely diversified trades have been given, to show that motion study is a problem of the most vital importance to the world. Some day an intelligent nation will awake to the fact that by scientifically studying the motions in its trades it will obtain the industrial supremacy of the world. We hope that that nation will be the United States. Already rated as the most progressive nation the world has ever seen, it will take a position far in advance of all, once it begins to give its earnest attention to this subject. Certain it is, that if we do not some other people will, and our boasted progress and supremacy will then be but a memory.

When one looks about him and sees the wasted time and money in every walk of life from useless motions, the mind becomes weary in contemplating the magnitude of the task. The bricklayer, the carpenter, the machinist, the shoveller, the clerk, even the editor in writing with his pen, make twenty motions where one would suffice. The actual wealth of the nation is in what it takes from the ground in the shape of crops or minerals plus the value added to these products by processes of manufacture. If by reducing the number of motions in any of these processes we can increase many fold the output of the worker, we have increased by that amount the wealth of the world; we have taken a long step in bringing the cost of living to a point where it will no longer be a burden to all but the very wealthy; and we have benefited mankind in untold ways.

Words fail the writer when he tries to express his appreciation of what Mr. Gilbreth has done in blazing a trail for future investigators. The work he outlines of investigating and reclassifying the trades by means of motion study is worthy of the brains of the most scientific investigators; it is worthy of the endowments of a Rockefeller or a Carnegie; it is worthy of the best efforts of the national government. Properly carried to its logical conclusion it would form the mightiest tool for the conservation of resources that the country could have. Our scientists could engage in no more important work than this.

ROBERT THURSTON KENT,
Editor Industrial Engineering.

MOTION STUDY

CHAPTER I

DESCRIPTION AND GENERAL OUTLINE OF MOTION STUDY

Necessity for Motion Study

Professor Nathaniel Southgate Shaler astounded the world when he called attention to the tremendous waste caused by the rain washing the fertile soil of the plowed ground to the brooks, to the rivers, and to the seas, there to be lost forever.

This waste is going on in the whole civilized world, and especially in our country. Professor Shaler's book, "Man and the Earth," was the real prime cause of the congress that met in Washington for the conservation of our natural resources. While Professor Shaler's book was right, and while the waste from the soil washing to the sea is a slow but sure national calamity, it is negligible compared with the loss each year due to wasteful motions made by the workers of our country. In fact, if the workers of this country were taught the possible economies of motion study, there would be a saving in labor beside which the cost of building and operating tremendous settling basins, and the transporting of this fertile soil back to the land from whence it came, would be insignificant. Besides, there would still be a surplus of labor more than large enough to develop every water power in the country, and build and maintain enough wind engines to supply the heat, light, and power wants of mankind.

There is no waste of any kind in the world that equals the waste from needless, ill-directed, and ineffective motions. When one realizes that in such a trade as bricklaying alone, the motions now adopted after careful study have already cut down the bricklayer's work more than two-thirds, it is possible to realize the amount of energy that is wasted by the workers of this country.

The census of 1900 showed 29,287,070 persons, ten years of age and over, as engaged in gainful occupations. There is no reason for not cutting down the waste motions in the vocations of the other almost half (49.7 per cent) of the population ten years of age and upward who do *not* engage in gainful occupations. The housekeepers, students, etc., on this list have as much need for motion saving as any one else, — though possibly the direct saving to the country would not be so great. But taking the case of the nearly thirty million workers cited above, it would be a conservative estimate that would call half their motions utterly wasted.

As for the various ways in which this waste might be utilized, that is a question which would be answered differently by each group of people to whom it might be put.

By motion study the earning capacity of the workman

can surely be more than doubled. Wherever motion study has been applied, the workman's output has been doubled. This will mean for every worker either more wages or more leisure.

But the most advisable way to utilize this gain is not a question which concerns us now. We have not yet reached the stage where the solving of that problem becomes a necessity — far from it! Our duty is to study the motions and to reduce them as rapidly as possible to standard sets of least in number, least in fatigue, yet most effective motions. This has not been done perfectly as yet for any branch of the industries. In fact, so far as we know, it has not, before this time, been scientifically attempted. It is this work, and the method of attack for undertaking it, which it is the aim of this book to explain.

Place of Motion Study in Scientific Management

Motion study as herein shown has a definite place in the evolution of scientific management not wholly appreciated by the casual reader.

Its value in cost reducing cannot be overestimated, and its usefulness in all three types of management — Military, or driver; Interim, or transitory; and Ultimate, or functional — is constant.

In increasing output by selecting and teaching each workman the best known method of performing his work, motion economy is all important. Through it, alone, when applied to unsystematized work, the output can be more than doubled, with no increase in cost.

When the Interim system takes up the work of standardizing the operations performed, motion study enables the time-study men to limit their work to the study of correct methods only. This is an immense saving in time, labor, and costs, as the methods studied comply, as nearly as is at that stage possible, with the standard methods that will be synthetically constructed after the time study has taken place.

Even when Ultimate system has finally been installed, and the scientifically timed elements are ready and at hand to be used by the instruction card man in determining the tasks, or schedules, the results of motion study serve as a collection of best methods of performing work that can be quickly and economically incorporated into instruction cards.

Motion study, as a means of increasing output under the military type of management, has consciously proved its usefulness on the work for the past twenty-five years. Its value as a permanent element for standardizing work and its important place in scientific management have been appreciated only since observing its standing among the laws of management given to the world by Mr. Frederick W. Taylor, that great conservator of scientific investigation, who has done more than all others toward reducing the problem of management to an exact science.

Vast Field for Motion Study

Now tremendous savings are possible in the work of everybody, — they are not for one class, they are not for the trades only; they are for the offices, the schools, the colleges, the stores, the households, and the farms. But the possibilities of benefits from motion study in the trades are particularly striking, because all trades, even at their present best, are badly bungled.

At first glance the problem of motion study seems an easy one. After careful investigation it is apt to seem too difficult and too large to attack. There is this to be said to encourage the student, however:

1. Study of one trade will aid in finding the result for all trades.

2. Work once done need never be done again. The final results will be standards.

Present Stage of Motion Study

We stand at present in the first stage of motion study, *i.e.*, the stage of discovering and classifying the best practice. This is the stage of analysis.

The following are the steps to be taken in the analysis:

1. Reduce *present* practice to writing.
2. Enumerate motions used.
3. Enumerate variables which affect each motion.
4. Reduce *best* practice to writing.
5. Enumerate motions used.
6. Enumerate variables which affect each motion.

Variables

Every element that makes up or affects the amount of work that the worker is able to turn out must be considered separately; but the variables which must be studied in analyzing any motion, group themselves naturally into some such divisions as the following:

I. *Variables of the Worker.*

1. Anatomy.
2. Brawn.
3. Contentment.
4. Creed.
5. Earning Power.
6. Experience.
7. Fatigue.
8. Habits.
9. Health.
10. Mode of living.
11. Nutrition.
12. Size.
13. Skill.
14. Temperament.
15. Training.

II. *Variables of the Surroundings, Equipment, and Tools.*

1. Appliances.
2. Clothes.
3. Colors.
4. Entertainment, music, reading, etc.
5. Heating, Cooling, Ventilating.
6. Lighting.

7. Quality of material.
8. Reward and punishment.
9. Size of unit moved.
10. Special fatigue-eliminating devices.
11. Surroundings.
12. Tools.
13. Union rules.
14. Weight of unit moved.

III. *Variables of the Motion.*

1. Acceleration.
2. Automaticity.
3. Combination with other motions and sequence.
4. Cost.
5. Direction.
6. Effectiveness.
7. Foot-pounds of work accomplished.
8. Inertia and momentum overcome.
9. Length.
10. Necessity.
11. Path.
12. "Play for position."
13. Speed.

In taking up the analysis of any problem of motion reduction we first consider each variable on the list separately, to see if it is an element of our problem.

Our discussion of these variables must of necessity be incomplete, as the subject is too large to be investigated thoroughly by any one student. Moreover, the nature of our work is such that only investigations can be made as show *immediate* results for increasing outputs or reducing unit costs.

The nature of any variable can be most clearly shown by citing a case where it appears and is of importance. But it is obviously impossible in a discussion such as this to attempt fully to illustrate each separate variable even of our incomplete list.

Most of our illustrations are drawn from bricklaying. We have applied motion study to our office and field forces, and to many of the trades, but our results on bricklaying are the most interesting, because it is the oldest mechanical trade there is. It has passed through all the eras of history, it has been practiced by nations barbarous and civilized, and was therefore in a condition supposed to be perfection before we applied motion study to it, and revolutionized it.

Since first writing these articles for *Industrial Engineering* it has been of great interest to the writer to learn of the conscious and successful application of the principles involved to the particular fields of work that have interested various readers. It was thought that unity might be lent to the argument by choosing the illustrations given from one field. The reader will probably find himself more successful in estimating the value of the underlying laws by translating the illustrations into his own vocabulary, — by *thinking* in his own chosen material.

The practical value of a study such as this aims to be will be increased many fold by coöperation in application and illustration. The variables, at best an incomplete framework, take on form and personality when so considered.

CHAPTER II

VARIABLES OF THE WORKER

Anatomy

A careful study of the anatomy of the worker will enable one to adapt his work, surroundings, equipment, and tools to him. This will decrease the number of motions he must make, and make the necessary motions shorter and less fatiguing.

Examples. — 1. If the bricklayer is left-handed the relative position of the pile of packs to the mortar box is reversed.

2. The staging is erected so that the uprights will be out of the bricklayer's way whenever reaching for brick and mortar at the same time. (See Fig. 1.)

3. Packs can be piled at a height with reference to the height of the mortar box that will enable stock to be picked up more easily by bending over sideways than by bending forwards. This latter case is, of course, on work where the non-stooping scaffold is not used.

4. The planks on the bricklayer's platform of the non-stooping scaffold, if made of two unconnected planks, will enable the bricklayer to lean either toward the stock platform or toward the wall without any other effort than that

of throwing his weight on one foot or the other, taking advantage of the spring of the planks. (See Fig. 2.)

5. The inside plank of the bricklayer's platform must extend in under the stock platform, or the bricklayer's leg

Fig. 1.—Non-stooping scaffold designed so that uprights are out of the bricklayer's way whenever reaching for brick and mortar at the same time.

will strike the edge of the plank of the stock platform when he reaches for stock.

6. The stock platform must not be wider than the minimum width that will permit holding the packets, or the lower-priced packet man will not place the packs exactly in that position that will require the least amount of straining of the high-priced workman, the bricklayer.

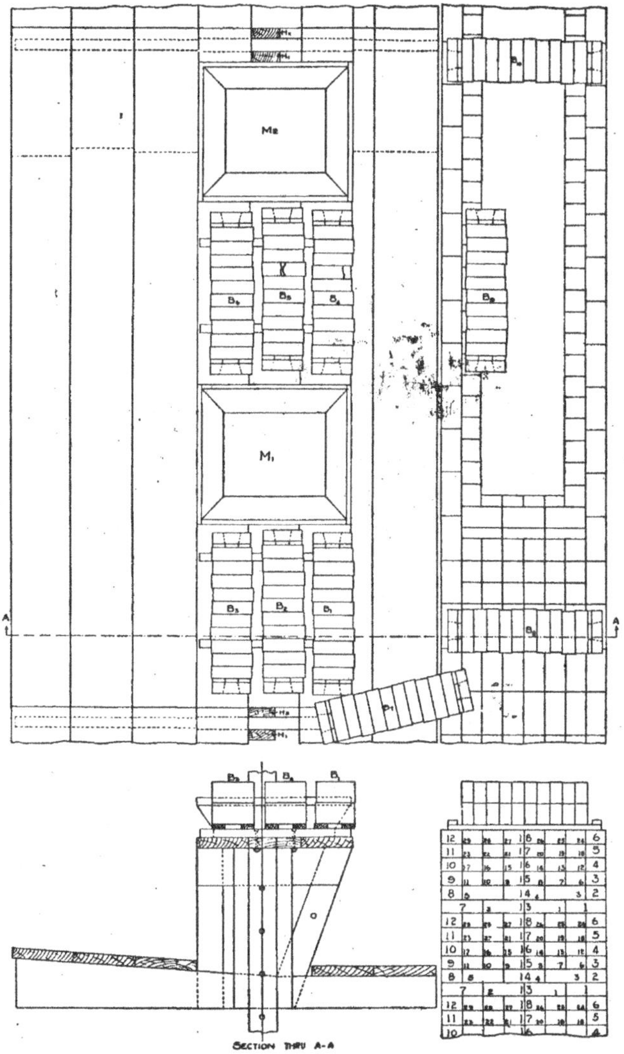

Fig. 2. — Gilbreth's patent non-stooping scaffold for bricklaying.

The numbers show the correct sequence of courses and tiers as laid from the non-stooping scaffold for the fewest, shortest, and most economical motions under the "Pack-on-the-wall" method.

Brawn

Workmen vary widely as to their brawn and strength.

When the actual work is being done, due consideration should be given to the percentage of efficiency that the men available possess. But all calculations should be made on the basis of using first-class men only. All data should be gathered from observations on first-class men only. In fact, so-called first-class men are not good enough. The best man obtainable anywhere is the best for observation purposes. The data gathered on that best man will then be considered as 100-per-cent quality. The men finally used can then be considered as of a certain percentage of perfect quality, and it should then be the aim of the management to attain 100-per-cent quality. This is one of the most important factors in the success of intensive management. The manager who wins is the one who has the men best suited for the purpose. Intensive management must not only recognize quickly the first-class man, but must also attract first-class men.

Everybody concedes that the size of the output depends, first of all, on the quality of the men.

Example. — We have found that a first-class laborer, if his work is so arranged that he does not have to stoop over, but can do his work with a straight back, can handle ninety pounds of brick on a packet (see Fig. 3) day after day and keep in first-class physical condition, while laborers of a class that does not have the right food cannot handle continuously over sixty to seventy pounds of bricks on a packet.

It is obviously better to have all one class of men, so that all instruction cards will be as nearly alike as possible. The size of the shovel, the weight of the hammer, the number of brick on the packet — these are variables that must

Fig. 3.— Arrangement of gravity conveyor for unloading brick so that laborer does not have to stoop.

also be considered when making out the instruction card — and these are all influenced by the brawn of the worker.

Contentment

Contentment affects the output of the worker. If he is contented, he will have his mind on his work, and he will be more willing to carry out the motions exactly as directed on the instruction card.

The contented worker does not require so large a percentage of rest for overcoming fatigue from his intensive efforts.

Contentment makes for loyalty to the management, for coöperating for maintainment of the best conditions, and for the protection and preservation of the property of the employer.

Creed

The term "creed" is used to cover religion, nationality, etc., — everything that might act as a bond of sympathy between workers and the people with whom they come in contact. On work where the output of each man is recorded separately, the question as to whether the creed of the workman is the same as that of his foreman, or superintendent, or employer, is of little consequence.

In places where the output of each man is not recorded separately, it is a recognized fact that instructions of the foreman or employer will be more apt to be carried out where there is a bond of sympathy between the employees, the foreman, and the employers. A bond of sympathy between the workman and the people who are to occupy the edifice upon which they are working will also increase the output.

The motions of a bricklayer working upon the wall of a church differing from his own religion are often vastly different from those that he is careful to make when the congregation to occupy it coincides with his belief.

In planning athletic contests also, it is well to group men according to their affiliations.

Example. — On engine beds and similar work, where the pieces are isolated, assigning gangs of men of different nationalities to the different beds will create extra interest in the contests. If this is not feasible, put the tall men on one bed and the short men on the other, or the single men against the married men, or eastern "pick-and-dip" men against western "string-mortar" men.

Earning Power

The matter of classifying men by their relative earning power is as important as classifying them by their relative brawn. It is better, of course, to have men as nearly as possible of one class only, and that the best class. Classing men by their earning power simplifies the work of the planning department in many ways. It enables it to prescribe the same motions to the entire class of men, to place them all under nearly the same conditions, to prescribe the same tools and surroundings, to place them together, and, finally, to have an athletic contest between the men of the same class.

Furthermore, the motions to be made are often entirely different for workmen of different earning power.

Examples. — 1. With masons and laborers of low earning power it is sometimes advisable to place the brick on the packets any way that will give the fewest motions for loading the packets, and to let the bricklayers lay them with their customary numerous motions, until men of higher earning power may be obtained to take their places.

2. With bricklayers and laborers of high earning power it is better to have the laborers pile the brick upon the packets so that the brick will be in that position that requires the least amount of motions of the bricklayer to pick them up and to lay them.

It is obvious that all motions performed in handling or transporting material before the material is used, cut up, or fabricated, should, theoretically, be performed by low-priced men, and that the work done by the high-priced

Fig. 4.—Right way to pick up brick from upper tier on the packet.

men should be limited as far as possible to the work of permanent character. As an example of this, the carrying of the brick and mortar to the scaffold is done by the mason's helper, while the carrying of the brick from the packet to its final resting place in the wall is done by the mason.

This same principle can be carried much further in all

trades than is usually customary to-day. For example, we have found that piling the brick face up and with the top side nearest the palm of the bricklayer's hand when his arm hangs in a natural position will save an average of one motion of the high-priced bricklayer per brick. (See Figs. 4, 5, 6, and 7.)

We have found a great increase in the number of brick it is possible to lay, and a decrease in the cost of laying them if the brick are placed by the low-priced man in the nearest practicable place in feet and inches from the place where they will finally rest in the wall. Not only this,

Fig. 5.—Wrong way to pick up brick from upper tier on the packet.

but the receptacle must be left with the material on it, so that the higher-priced man can lift the receptacle and its contents simultaneously at the exact time the materials are wanted to a place still nearer to the place where the material will be finally used, to be transported from there to their final resting place by a still higher-priced man. (See Figs. 2 and 8.)

This use of "low-priced men" does not mean the use of mediocre men. The men used, of whatever price, should be the best men of that class obtainable.

Experience

That previous experience is an element to be considered is obvious. This fact is so well recognized that the expres-

Fig. 6. — Right way to pick up brick from the lower tier on the packet.

sion "You can't teach an old dog new tricks" may be heard around the world. While this may be true with dogs, it is not true with workmen. On a short job it may not be advisable to attempt to change radically the lifetime customs of a local workman. But recording the output of each man separately will tell whether or not it is advisable to make out the instruction card in accordance with the previous experience of the workman, or in accordance with the way in which actual records have proved to be productive of the highest outputs. Experience varies

Fig. 7.—Wrong way to pick up brick from the lower tier on the packet.

widely, and the habits formed are often difficult to overcome.

Example. — A bricklayer from certain sections of New England has been accustomed to pick up mortar with a trowel at the same time that he picks up brick with the

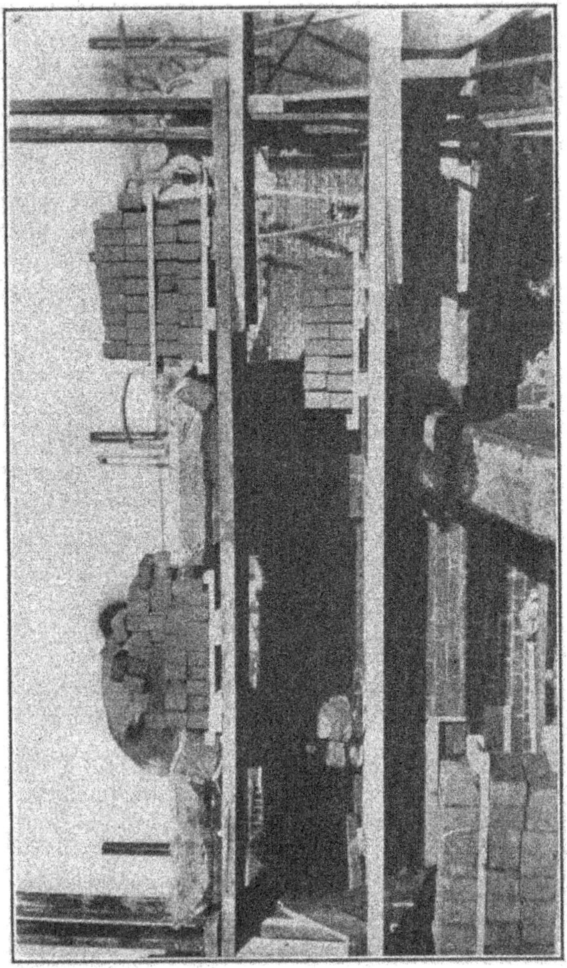

Fig. 8. — The non-stooping scaffold for handling packs of brick with the fewest and shortest motions.

other hand. This is called the "pick-and-dip method." The size and shape of his mortar receptacle, the arrangement of the brick and mortar on his scaffold, the shape of the scaffold itself, the sequence in which he builds the vertical tiers and the horizontal courses, and, finally, the labor-union rules themselves, are fashioned after the consequences of using a small trowel, just large enough to pick up sufficient mortar for one brick only.

Fig. 9.—The usual method of providing the bricklayer with material.

A bricklayer so trained finds it difficult at first to adapt himself to the "string mortar" method of the West. The western-taught bricklayer experiences the same difficulties in adapting himself to the "pick-and-dip" method with the speed of the eastern bricklayer. But their difficulties are nothing compared with those that the employer experiences who puts the good points of both systems on any one job.

Not only do habitual motions become fixed, but also the previous experience of the bricklayer is often the cause of his making *too many motions, i.e.*, unnecessary motions. He seldom, if ever, has been rigidly trained to use a certain number of definite motions. It takes time and patience to induce him to adopt a standard method.

On a small job it is advisable to select those men for the leads and the trigs who are best fitted to be leaders, that is, who are best prepared by previous experience to carry out without delay the requirements of the instruction cards — but give due consideration to the previous experience and habits of work of the workmen.

On a large job, however, it is most economical to insist on standard methods and standard motions that will produce the highest outputs, without regard to the previous training of the workmen. Attract and retain those workmen who can follow out their instruction card and as a result produce the high records of outputs.

FATIGUE

Fatigue is an important variable to consider when selecting those motions that will give the most economy and that make the "standard motions." It goes without saying that the motions that cause the least fatigue are the most desirable, other things being equal.

Fatigue is due to a secretion in the blood.

To quote from an article signed "I. M. T." in the *American Magazine* for February, 1910:

"The toxin of fatigue is the phrase the physicians have given us with which to jar the attention of those who can only be stirred by harsh words. It has been demonstrated in the last few years that fatigue is due to an actual poison not unlike the poison or toxin of diphtheria. It is generated in the body by labor. But the system takes care of itself and generates enough anti-toxin to take care of a normal amount of toxin or poison. If it continues to be produced in abnormal quantities the system cannot grapple with it. There is a steady poisoning of the body, with all the baneful effects, mental and moral, as well as physical, that poison produces."

Continuous hard work, however, like proper training, puts the body into that condition that best overcomes fatigue.

Fatigue is due to three causes:

1. Fatigue due to coming to work improperly rested (fatigue brought to the job).

2. Unnecessary fatigue, due to unnecessary work, unnecessary motions, or uncomfortable positions, surroundings, and conditions of working.

3. Necessary fatigue, due to output.

Every motion causes fatigue. The same motions in the same trade cause about the same fatigue for all first-class men, and they all require about the same amount of rest to overcome fatigue, provided their habits and mode of living are the same outside of working hours.

The amount of fatigue caused and the percentage of rest required in many different kinds of work have been computed by Frederick W. Taylor with great exactness. He has assigned the various workers to classes and accurately computed the "task" from his records.

We have no such records as Mr. Taylor has gathered, but we have numerous records of outputs of different men on several kinds of work. We know that the amount of rest actually required by a workman increases with the discomfort of the position in which he works. We also know that the speed, hence the output of the worker, decreases rapidly if there is much fatigue to overcome.

Example. — A bricklayer can lay brick for a few minutes quite as quickly when he picks up the brick from the level of the platform on which he stands (see Fig. 9), as he can when he picks up the brick from a bench twenty-four inches above the level of the platform on which he stands (see Figs. 10, 11, and 12), but he cannot keep that speed up, because he requires more rest to overcome the greater fatigue.

It is not simply for the welfare alone, although that reason should be sufficient, but for economic reasons as well, that the men should be so placed and equipped that their work is done under the most comfortable conditions.

Examples. — 1. It is a recognized fact that a cluttered-up floor under a workman's feet will tire him quite as much as the productive work that he is doing. A

Fig. 10. — Standard practice of providing bricklayers with materials — mortar boxes on the floor and low piles of packs when the wall is below the level of the top of the floor.

Fig. 11. — Standard practice of providing bricklayers with materials — raised mortar boxes and high piles of packs when the wall is waist high.

smooth-planked floor will enable a bricklayer to lay many more brick than will earth that has been leveled off.

2. A bricklayer can stoop over and pick up anything from the floor with one hand with much less fatigue if he has a place to rest his other hand while he is stooping,

Fig. 12.—Standard practice for providing bricklayers with material for pick-and-dip method of bricklaying when the wall is shoulder high.

because he puts his weight on one foot and lifts his other foot out behind him, which does not tire the muscles of his back nearly so much.

Slow motions do not necessarily cause less fatigue than quick motions, and, per unit of work done, may cause much more fatigue than quick motions.

The amount of work done per motion may not be fatiguing proportionately to the size of the unit.

Example. — Lifting ninety pounds of brick on a packet to the wall will fatigue a bricklayer much less than handling the same number of brick one or two at a time. Consequently with the same amount of fatigue the workman will handle several times as many brick on packets as he can handle one or two at a time.

We have, then, under this variable two tasks to perform:

1. To eliminate unnecessary fatigue. This we do by studying and fixing the variables; that is, by standardizing the work.

2. To provide for rest from necessary fatigue, and to utilize rest time.

Under old forms of management workmen "should keep busy at something," even if prevented from doing their regular work. An idle workman was considered a disgrace. The consequence of this was that the workman took his rest while working, or made believe work while resting. The old-fashioned kind of rest is called "systematic soldiering." It is the curse of the military type of management. It is a form of cheating that has been made respectable by the conditions forced upon the workers by the employers.

Under scientific management the evils of soldiering are eliminated, and the correct definite percentage of rest required is recognized and provided for. When a man is prevented by causes beyond his control from doing his regularly assigned work, he is told to use the opportunity for rest,—not to take such rest as can be obtained by making slow and useless motions, that will give him an

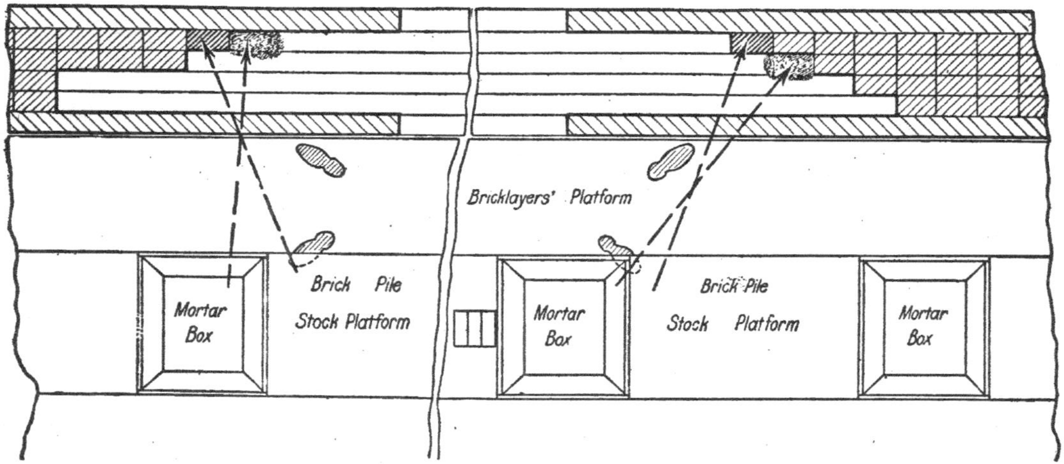

Fig. 13. — Pick-and-dip method of bricklaying on the interior or filling tiers.

Working from left to right. — The mortar should be laid to the right of the brick and deposited on the wall at exactly the same instant that the brick is deposited on the previous trowelful. Otherwise the brick would wait until the mortar was deposited.

Working from right to left. — The mortar and brick are deposited in different tiers, so that they may be both deposited at the same time. The apprentice must be taught on his first day the right positions for his feet and the right motions for the greatest economy.

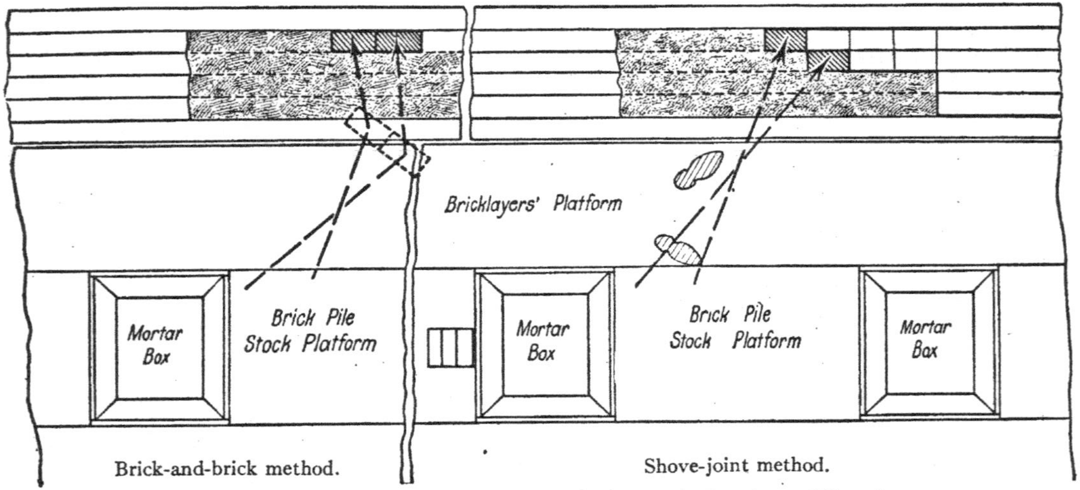

Brick-and-brick method. Shove-joint method.

Fig. 14.—Stringing-mortar method of bricklaying on the interior or filling tiers.

Brick-and-brick method of bricklaying, workirg from right to left.—In the brick-and-brick method the brick are picked up simultaneously, put together end to end in the air, and guided as one brick to place for the greatest speed. This method is not used when shove joints are desired.

Shove-joint method of bricklaying, working right to left.—The two brick are deposited in different tiers at the same time.

industrious appearance to the casual observer, but to rest, the 100-per-cent kind of rest.

There are cases where chairs and reading tables have been provided with beneficial effect for workers to occupy when delayed for a few minutes. They get the rest, and their presence at the table acts as a danger signal to the management.

When a man is fatigued to the point where it is impossible for him to do his best work he should be made to rest. He must not do anything but rest until he is in that condition that will enable him to fly at his work and perform it with the fastest standard motions possible.

Rest does not necessarily mean idleness. The worker can spend the rest period reading his instruction card, or filling out his record of output on the card, or in some other form of restful work. A change of work is often a rest. By performing the above two tasks well, we secure the greatest output per day and the fewest hours per day without injury to the health of the men.

Habits

The habits of the workman have much to do with his success in eliminating unnecessary motions and in adopting quickly and permanently standard methods. The term "habits," as here used, includes not only personal "habits," so-called, but also habits of thinking, habits of working, etc.

Habits brought to the work may act as a deterrent or

as an aid to its best performance. They embrace a group of sub-variables which are difficult to describe and analyze, and are of immense importance in influencing output.

That acquiring good habits of work makes the worker more versatile as well as more efficient is forcefully stated by Mr. Gantt in his book on "Work, Wages, and Profits." He says:

"The habits that a man has to acquire to become efficient in one class of work stand him in good stead in becoming efficient in other work. These habits of work are vastly more important than the work itself, for it is our experience that a man who has become efficient in one thing, readily learns to become efficient at doing other things."

Health

The health of the worker may be affected by:

1. Other things than his work and the conditions under which it is done.

2. The work.

Consideration of other things than the work may properly be left to the welfare department. This department can most successfully define the scope of its work by attempting to improve the man himself and his surroundings in every way that will make him a better and more successful worker. This criterion will satisfy both employer and employee as to the appropriateness, justness, and utility of the work of the welfare department.

The life of the man when away from work is only in so far subject to the inspection and jurisdiction of the so-called "welfare" department as that department can show itself able to make of the man a more valuable economic unit to himself and to the community.

If the welfare department makes an efficient workman the product of its work, the philanthropic by-products will take care of themselves.

The work itself should be laid out in such a way that its performance will add to and not subtract from health. A proper study and determination of the variables that affect the surroundings and the motion will go far to insure this. Moreover, standardized work will transform the workman.

Henry L. Gantt, in a most stimulating paper on "Training the Workmen in Habits of Industry and Coöperation," read before the American Society of Mechanical Engineers, December, 1908, says of workmen:

"As they become more skilled, they form better habits of work, lose less time, and become more reliable. Their health improves, and the improvement in their general appearance is very marked. This improvement in health seems to be due to a more regular and active life, combined with a greater interest in their work, for it is a well-known fact that work in which we are interested and which holds our attention without any effort on our part, tires us much less than that we have to force ourselves to do."

This Mr. Gantt says in speaking of the benefits of the "task and bonus" system; but the same thing is undoubtedly true of men working under standards derived from motion study.

Mode of Living

Mode of living has been more or less touched upon under "health" and "habits." It is a complex variable, difficult to analyze and difficult to control. Its effects on output are for this reason all the more far-reaching and demand scientific investigation.

Nutrition

This is a subject that has been investigated much more scientifically with regard to horses and mules than with regard to workmen, but cases are seen on every hand where it is more profitable to furnish the most nutritious food to the men gratis than to permit them to have the usual poor food of the padrones' storehouse. In the building of a new town in Maine it was found to be economical to spend considerable sums of money for supplying food for the men at less than cost, rather than to have them eat the food provided by the local boarding houses. The nutritive value of various foods and the amount of energy which various diets enable one to put forth have been made a study in training soldiers. There must be many data available on the subject, and the government should collect them and issue a bulletin for the use of the welfare departments of large employing organizations. The army might also serve as an example in many other ways to the student of economics. The "Tactics" are admirable "instruction cards," conforming to many of the laws of motion study. It seems unfortunate that the governments of the world up to the present time have confined all of their attempts to standardize motions to the arts of war, and have done nothing in this line in the arts of peace.

Size

Size of men, with relation to their motions, has much more influence than is usually realized.

Short men are usually the best shovelers where the shovelful need not be raised much in doing the work, such as in mixing mortar and concrete. Few foremen realize that this is because a short man does fewer foot-pounds of work in doing the same amount of shoveling. On the other hand, when men are shoveling in a trench, the taller the men, usually, the more the output per man.

Oftentimes a staging is built at a height below a set of floor beams that enables the men to work to best advantage. On such a staging men should be selected of as nearly the same height as possible.

Skill

The workman with the most skill is usually the one who can adapt himself quickest to new methods and conditions.

Example. — A bricklayer who has great skill in his trade can instantly lay a brick in the same manner that he is once shown. To get him to do so constantly when not supervised is difficult, but that can be quickest impressed upon his mind if he is shown the reason for every change demanded of him.

To make sure that the worker of the future acquires his skill properly, is the most important task here. This can be done only by insisting continuously on conformity to scientifically derived standards from the beginning of his training.

Example. — The best results from a motion-study standpoint can be attained only by teaching the apprentice from his first day to lay the brick with the standard motions regardless of the looks of the work. If the work is not good enough to permit the brick to remain on the wall, a skilled bricklayer should fix it, until the apprentice can lay the brick with the prescribed standard motions in a manner good enough to permit the work to remain as a part of the structure.

The apprentice should not be permitted to depart from the standard motions in any case until he has first acquired them as a fixed habit. The most pernicious practice is the generally accepted one of first having an apprentice do perfect work and then attempting to make speed later. The right motions should be taught first, and the work taken down and rebuilt until it is up to standard quality. This is the *only* way to get the full benefits of the economics of motion study. (See Figs. 13 and 14.)

The workman who will make the highest outputs of the future will be he who has as a *habit* those standard motions that are the most productive when operated under standard conditions.

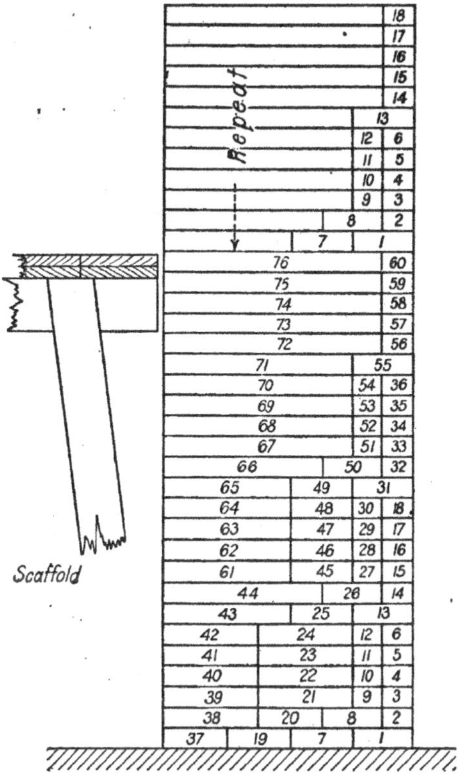

Fig. 15.—The numbers show the correct sequence of courses and tiers as laid from the ordinary scaffold for the fewest, shortest, and most economical motions.

TEMPERAMENT

The temperament of the man has more to do with the motion he uses than one usually supposes.

Example.—Many expert face bricklayers would quit a job rather than lay common brick on interior walls, even

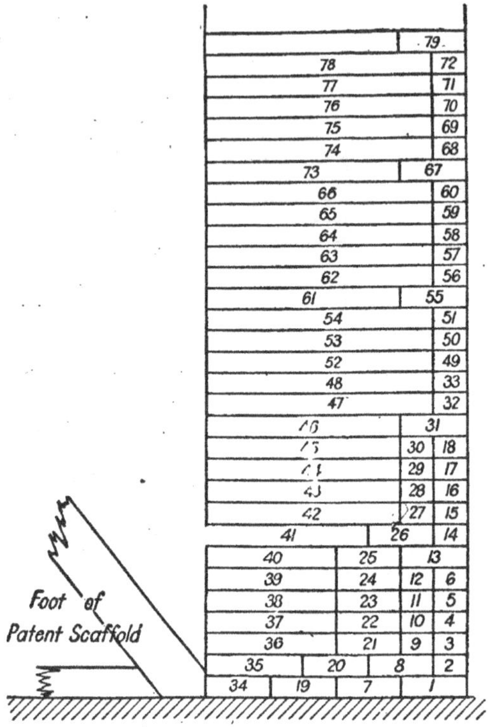

Fig. 16. — The numbers show the correct sequence of courses and tiers as laid from the non-stooping scaffold for the fewest, shortest, and most economical motions. This scaffold permits a much larger percentage of the brick to be laid as "filling in," instead of as "backing up," consequently requiring less skillful motions.

though they might earn higher wages on the inside work. Other bricklayers prefer to lay common brickwork, — not that they doubt their ability to lay the face brick, but because they like the strenuous athletic contests for high scores of output and high pay. To them there is no monotony in laying common brick day after day, for to the skilled mason brick are not so nearly alike as are human beings.

A bricklayer interested in his work will often remember the characteristics of one certain brick years after he has forgotten the wall upon which it was laid.

Therefore the temperament of the man must be taken into consideration when placing the men. When they are best placed they follow their instructions on the subject of motion, and higher scores will be the result.

Training

"Training" is so closely related to "skill" and "experience" that it is difficult to separate it from them. We use the word to mean both the worker's theoretical and practical equipment for his work, his entire preparation. The problem is to see that the worker has both kinds of equipment, acquired in the most useful, balanced method possible.

The training of the available worker must always be considered in estimating the time that it will take him to acquire standard methods and the output that can be expected of him. The training of the worker of the future should be planned to fit him for standard work. The training of the apprentice on the work to-day is usually defective because he has little or no training in theory at the same time that he is getting his practice. Furthermore, the journeyman who is his instructor not only has had no training in pedagogy, but often lacks the benefits of the elements of a common-school education. The usual time of apprenticeship in the building trades in this country is three years, or until the apprentice is twenty-one years old.

On the other hand, the boy taught in the trade school lacks training under actual working conditions. The question of dollars and cents to make for the employer, special fitting for high wages for himself, and the knowledge of the principles underlying the requirements necessary in order to obtain specially high outputs from intensive management, are wholly lacking.

The present apprenticeship system is pitiful and criminal from the apprentice's standpoint, ridiculous from a modern system standpoint, and there is no word that describes its wastefulness from an economic standpoint.

Summary

Before turning to the variables of the surroundings, it may be well to summarize. The variables of the worker consist of the elements of the equipment that the worker brings to his work, both those that he was born with and those that he has acquired. These are mental and physical.

We have concluded:

1. That first-class men should always be secured if that be possible.

2. That everything possible should be done to preserve

and to add to the natural powers and capacities that the worker brings to his work.

3. That standard practice derived from motion study does add to the natural powers of the worker, and both shortens his hours of work and adds to his output.

4. That training based on the laws underlying standard practice will enable the worker of the future to attain still higher efficiency and output.

CHAPTER III

VARIABLES OF THE SURROUNDINGS

We turn now to the variables of the surroundings. These differ from the variables of the worker in that we can influence them more quickly and more directly. In discussing the variables of the worker, we deal more or less with the past and the future. The variables of the surroundings are each and all distinctly of the present.

Appliances

The "standard conditions" maintained by the employer are a most important factor for high outputs. It is obvious that the appliances furnished the workman and the motions used are interdependent on each other.

Examples. — 1. The bricklayer could not be expected to pick up the brick so that he would not have to spin or flop it in his hand unless it were delivered to him in the right position on a packet.

2. The bricklayer could not be expected to have so high an output if he had to stoop over in order to pick up his stock as he would have to do if the scaffold did not have a bench that obviated bending.

3. The bricklayer could not be expected to lay brick without turning around or bending over unless he was provided with packs of bricks that could be lifted bodily

and placed upon the wall in units as large as could be economically handled.

4. The bricklayer could not be expected to do away with those motions that are necessary to remove the lumps from under a brick if there were holes in the sand screen and no pug mill to break up the lumps.

It is most important that the workman should be given "handy conditions" under which to work, that is, the "most comfortable," or those that require the "least percentage of rest" to overcome fatigue.

Examples. — 1. The bricklayer must obviously have a scaffold to stand upon that permits adjusting the height of the platform on which he stands to a standard distance below the top of the ever-growing wall on which he is laying the brick. We have found that the best height is from twenty-four to thirty-two inches below the top of the wall. If the wall is being laid overhanded, the height should not be over twenty-four inches, while if the wall is not being laid overhanded, thirty-two inches is the better height.

It is obvious that the bench from which the stock is picked up should be maintained at a standard distance above the platform on which the man stands. Also the platform on which the laborer walks should be located at the standard distance below the stock platform that will enable him to deposit the brick and mortar in a manner that will cause the least fatigue. Therefore, the three platforms for bricklayer, stock, and tender should be fixed with relation to one another, and movable in relation to the top of the wall, capable of being hoisted as the wall grows without stopping or disturbing the men.

2. The elevator for hoisting the brick and mortar should always be arranged so that it can, when desired, land above the top of a staged wall, and thus the brick and mortar can be wheeled down to the scaffold on the floor below. Then the tenders can wheel down with full loads and wheel the empty barrows up to the floor above.

3. Make a table, barrel, or box to put near the workman, no matter what his trade is, so that he will not have to stoop over and pick up his tools. Provide something to lean his shovel against or to hang his shovel on when he is alternately shoveling and wheeling to cut down time and to reduce the fatigue of stooping over and picking up the shovel.

The motions to be used and to be avoided are largely determined and affected by the appliances used; therefore for the highest outputs the right appliances must be devised, standardized, used, and maintained, otherwise the motions cannot be standardized. Furthermore, it is much easier to standardize motions with standard appliances than without them.

Clothes

The clothes that the workman wears may be a hindrance or a help to him in his work. Tight or ill-fitting clothing may restrict motions. Fear of ruining clothing may seriously cut down the speed of the worker.

On the other hand, clothing designed and specially adapted to the work that the worker has to do may increase output to a surprising extent.

Not till the advantages have been appreciated of having working clothes made the subject of study from the motion-economy standpoint will manufacturers provide the garments needed. But they are only too anxious to meet every demand as soon as they are conscious of it. Once let the specialized clothes for the worker be standardized and they will be placed immediately upon the market in inexpensive, durable, and attractive shape.

As for their reception by the worker, as soon as he realizes that they increase his efficiency, and are a badge of specialization and not of servitude, he will be ready and glad to welcome them.

Color

The stimulating effect of color upon workers is a subject to be investigated by psychologists. The results of their study should be of great benefit, especially to indoor workers. Motions could undoubtedly be made simpler by the proper selection of the color of painting and lighting in the workroom.

In our work we have to deal chiefly with color as a saver of motions. Color can be seen quicker than shape. Therefore, distinguishing things by their color is quicker than distinguishing them by the printing on them.

Examples. — 1. The various pipes in a pipe gallery

can best be recognized by painting them different colors.

2. The right-hand end of the packet is painted black, in order that when carried in the right hand of the laborer it can be placed so that the bricklayer can pick up each brick without spinning or flopping the brick in his hand.

3. Painting tools different colors, and also the place where they are to be placed in the drawer or the chest the same color, saves motions and time of motions when putting them away and finding them next time.

4. When low-priced men bring packages of any kind to higher-priced men to use or handle, the packages should always be painted, stenciled, or labeled with a distinguishing color on one end and on top. This will enable the low-priced workman to place the package in the manner called for on the instruction card with the least thought, delay, and motions. It will also enable the high-priced man to handle the package with no such lost motions as turning the package around or over.

5. Oftentimes the workmen who are best fitted physically for their work cannot read, or at least cannot read English. Even if they could, it would take some time to read the stenciled directions on the non-stooping scaffold to the effect that "this side goes against the brick wall." It will greatly reduce the number of motions to paint the side that goes next to the wall a different color from the side that goes away from the wall.

Entertainment

Music. — The inspiring and stimulating effect of music has been recognized from ancient times, as is shown by the military band, the fife and drum corps, the bagpipe of the Scotchman, down to the band that rushes the athlete around the track or across the field.

The singing of gangs at certain kinds of work, the rhythmic orders that a leader of a gang shouts to his men, and the grunting in unison of the hand drillers, show the unifying as well as the motion-stimulating effect of music and rhythm.

That some of the trades can have their motions affected in time and speed by music, to a point that will materially affect the size of their outputs, is a recognized fact.

Some of the silent trades have used phonography and musical instruments to entertain the men while they were working. It was found it paid the employer to furnish stimulating records at his own expense, so that the workmen would make more and quicker motions, rather than to permit the employees to furnish phonographic records at random at their own expense.

Reading. — Reading as a stimulus to output has been used with excellent results among the cigar makers.

It is also interesting to read in an article on "Three Months in Peonage" in the March, 1910, issue of the *American Magazine*, that story-telling may produce the same good results.

"The four packers under me," says the writer, a German white, who was working with peons at packing tobacco in Mexico, 'knew no greater joy than to listen to a fairy tale with the regulation princess and dragon, and if I could but tell them one, or one of their number did so, the work went twice as fast, and they were happy."

The excellent and direct effects of entertainment upon health, fatigue, etc., are subjects for the scientist to study and the planning department and the welfare worker to apply. The effects of entertainment upon output should be studied by the student of motion economy. This variable alone furnishes a vast field for investigation.

Heating, Cooling, Ventilating

Heating, cooling, ventilating, and humidizing are closely allied, because all can be done with one and the same apparatus, and all greatly increase the workman's comfort, health, and possible number of motions.

Maintaining desired temperature in summer as well as winter by forcing into workrooms air that has been passed over heating or refrigerating coils has a great effect on the workman. Many factories, such as chocolate factories, have found that cooling the air for better results to the manufacturing process also enables the workers to produce more output — an output quite out of proportion to the cost of providing the air.

In many trades requiring great alertness and physical strength the proper heating and ventilating will allow the workman to dress in a costume specially adapted to his work, or to strip almost to the athlete's suit, with a consequent increased number and effectiveness of motions.

The degree of temperature and the percentage of humidity desired for each day of the year should be determined. The man in charge of the heating should receive no bonus for small consumption of fuel unless he also maintained the temperature and humidity called for on his instruction card.

The subjects of heating, ventilating, etc., are well covered by Mr. Hugo Diemer in his book on "Factory Organization and Administration." The proper time to consider these subjects is when the building is designed, but too often at that time the all-important question is, — How cheaply can the building be built? Ultimate saving will justify almost any conceivable first costs.

Lighting

The subject of lighting has, indirectly as well as directly, a great influence upon output and motions, as upon the comfort of the eye depends, to a large extent, the comfort of the whole body.

The arrangement of lighting in the average office, factory, or house is generally determined by putting in the least light necessary in order that the one who determined the location of the light may be able to see perfectly. This is wrong. The best light is the cheapest. By that is not meant that which gives the brightest light. In fact, the light itself is but a small part of the question. Go into any factory and examine every light, and you will notice that as a rule they are obviously wrong. A light to be right must pass five tests:

a. It must furnish the user sufficient light so that he can see.

b. It must be so placed that it does not cause the user's eyes to change the size of the diaphragm when ordinarily using the light.

c. It must be steady.

d. There shall not be any polished surfaces in its vicinity that will reflect an unnecessary bright spot anywhere that can be seen by the eyes of the worker.

e. It must be protected so that it does not shine in the eyes of some other worker.

The use of polished brass and nickel should be abandoned wherever it will shine in the worker's eye.

For work done on a flat surface, like the work of a bookkeeper or a reader, the light should be placed where the glare will reflect least in the worker's eyes; where the work is like the examining of single threads, the relative color and figured pattern of the background, as well as good light, is important. This is obvious. So is nearly everything else in good management. Go into the buildings among the workers, the students, and the scientists and see how rarely it is considered. All of this is not a question of getting the most out of the light. Light in a factory is the cheapest thing there is. It is wholly a question of fatigue of the worker. The best lighting conditions will reduce the percentage of time required for rest for overcoming fatigue. The difference between the cost of the best lighting and the poorest is nothing compared with the saving in money due to decreased time for rest period due to less fatigued eyes.

It is a similar case to the taxicab concerns — they charge their drivers with gasoline and tires and mileage, accidents, etc., but they furnish the lubricating oil free. The fallacy of the common practice of putting the lighting in the hands of the man whose merit is measured inversely as the coal bill is obvious.

The sub-variables involved make the problem as to exactly what lighting is most desirable difficult of solution. The proper solution will have such a beneficial effect, not only upon the man's work, but also upon his welfare, that no time or effort expended upon it can be too great.

Quality of Material

It is essential to the use of standard motions and the resulting large output that all material used shall be in exactly that state in which it can be most easily handled by the worker.

Examples. — 1. If there are lumps in the mortar, due to pieces of brick or shavings or lumps of lime, or cement or coarse pebbles in the sand, it is impossible for the bricklayer to do his best work.

2. If the sand is not selected with reference to the thick-

ness of joints, if the sequence of tiers and courses (see Figs. 15 and 16) and the thickness of joints is determined by the whim of the bricklayer on the lead, instead of by the planning department, it is out of the question to expect high outputs. On the other hand, if the material is of exactly that consistency with which it can be best handled, and the other conditions are determined on the instruction card, much better speed can be obtained.

3. When using cement mortar made of cement and sand and no lime, the bricklayer will do more and better work if a tender is kept on the stock platform tempering the mortar to just the right consistency for the bricklayers.

4. If the brick are all handled in packs on packets from the time that they arrive upon the job until they reach the bricklayer's hand, they will each be of better quality, due to there being little or no chipping from handling and throwing about. The bricklayer will then be saved the useless motions of picking up brick that are chipped and discarding them again, to be used only when laying in the filling tiers.

Rewards and Penalties

The stimulus that rewards and penalties give motions is obvious. The discussion of reward and punishment would come under the head of compensation. It must be left to the cost reducing system to determine just what system of compensation will induce the men to do their swiftest, best work.

Size of Unit Moved

The most advantageous size of unit to use is a difficult problem to solve, and is often controlled by some outside factor. For example, the most economical size of brick has been determined by the cost and other conditions relating to the making and baking, and not by the conditions of handling and laying. When the conditions of laying are studied scientifically, as they are to-day, one is forced to the conclusion that, for the greatest economy, the size of common brick should be changed materially from that of the present practice in America. The usual size of the brick used in England is much larger than the customary size used here.

It is obvious that there is some size of unit that is the most economical to make the standard package for handling brick in bulk. We have found it to be ninety-two pounds for a first-class laborer, either for piling or loading and unloading brick from carts. (See Figs. 17 and 18.)

Careful examination of brickwork with the object in view of selecting the most profitable motions has entirely revolutionized the methods of bricklaying. For example, the size of unit that is picked up when loose brick are handled must be one brick for each hand. The packet enables us to pick up about eighteen brick at once.

The fountain trowel permits us to pick up and carry to the wall and spread mortar for twenty-one brick at one time without dropping the regular trowel which forms a temporary handle to it. (See Fig. 19.)

Fig. 17. — Two-horse carts with horses changed from the empty to the full carts will require fewer and cheaper motions than any other methods of transportation.

Fig. 18. — Loading carts with 90-pound packs of brick fed on gravity conveyors.

The two-wheeled trucket permits carrying twelve packets, or 216 brick (see Fig. 20), while the hod carries 18 brick, and the one-wheeled barrow carries 60 loose brick.

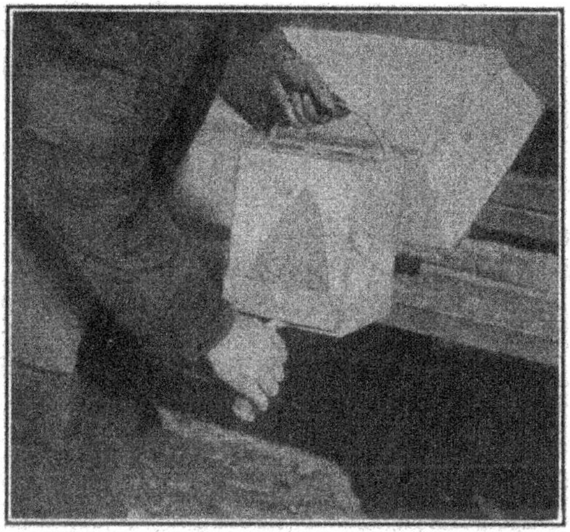

Fig. 19.—The Fountain Trowel for conveying and spreading mortar for 21 brick. The regular trowel forms a temporary handle for lifting and guiding it.

Special Fatigue-eliminating Devices

Only the careful student of management realizes how much the speed of the worker can be increased by providing him with all possible aids toward doing his work.

Mr. Fred. W. Taylor, in his paper on "Shop Management," tells of a study he made of overhauling a set of boilers.

"He [the writer] did all of the work of chipping, cleaning, and overhauling a set of boilers, and at the same time made a careful time study of each of the elements of the work. This time study showed that a great part of the

Fig. 20. — The two-wheeled trucket for carrying 12 packs. This trucket is so perfectly balanced that it causes less fatigue than the ordinary wheel-barrow.

time was lost owing to the constrained position of the workman. Thick pads were made to fasten to the elbows, knees, and hips; special tools and appliances were made for the various details of the work. . . . The whole scheme was much laughed at when it first went into use, but the trouble taken was fully justified, for the work was better done than ever before, and it cost only eleven dollars to completely overhaul a set of 300 horse-power boilers by this method, while the average cost of doing the same work on day work without an instruction card was sixty-two dollars."

In reading this, it must be remembered that the fatigue-eliminating devices were only one element in increasing speed and reducing costs. But, on the other hand, it must be remembered also what a large element they were in adding to the comfort and ultimate well-being of the worker.

SURROUNDINGS

"Surroundings" have been previously discussed under "Fatigue," "Appliances," etc. It is only necessary to say here that the surroundings of the worker should be standardized, the standard being derived from a study of all the variables.

It is obvious that the highest possible records of output cannot be obtained unless the workers are furnished with a standard instruction card made out by the best man obtainable, one who knows more about their work than they do, and who can, and does, provide them with standard conditions that fulfill the most economical conditions of motions. Even then daily outputs and unit costs must be watched, so as to take advantage of the slightest change of conditions that affect costs. In practice, the unit costs must always also include the wages of the recorder, otherwise one cannot tell when the wages of the recorders are not deceiving as to actual unit costs under this intensive management.

Tools

The influence of the tools used upon the output is large. No workman can possibly comply with standard motions unless he has the standard tools. No worker should ever be obliged to furnish his own tools, if large output is expected. When workmen are obliged to furnish their own tools (due to their having too much thrift, lack of money, or fear of having them stolen), they usually use one size only of the same kind of tool. On many kinds of work greater output can be obtained by using two or more sizes of a tool.

Example. — The bricklayer should use a smaller trowel on pressed brick and a larger trowel on common brick.

Again, where workmen furnish their own tools, they use them after they are too much worn. A shovel with a worn blade will require several motions to push it into the material to fill it. It is cheaper in this case to cut off the handle of the shovel, so that the men cannot use it. Where no records are kept of their individual outputs the men always choose the shovel with the small blade.

It is especially important that apprentices should be supplied with proper tools. According to the usual practice the apprentice is taught with any tool procurable. He becomes adept and skilled, but often becomes so accustomed to the poor tool he has used that he finds it difficult to adapt himself to the use of a better new tool. This seriously hinders his complying with demands for standard quantities of output.

Tools should be of standard size and pattern. Workmen should invariably be made to use a tool that will enable them to make standard-sized outputs instead of using a tool that may seem "handier" to them. You cannot expect a man to comply with standard motions unless he has the standard tool for which his standard instruction card was made out.

The customary method in the past for determining the best weight of tool to use was to guess at it, and to use that size of tool which was thought to be the "handiest," or which it seemed could be used with the least fatigue.

Makers of hand tools cater to the whims of the local workmen, and, as a result, hand tools are made of many different designs in different parts of the country. Makers spend and waste great sums of money making experiments and conducting selling campaigns of odd or new designs of tools that have no merit from a motion-economy standpoint. There should be a bureau of testing, where the actual value of new shapes, designs, and sizes of tools could be tested and rated in percentages of efficiency from the standpoint of motion study.

Critics will say that such a scheme will crowd out new designs, and the benefit of the individual's inventions will be lost. But it would not; on the contrary, the testing would give great stimulus to inventors, designers, and tool makers, for they could then obtain the immediate attention of the buyers, because they would have the standard stamp of merit that comes from the record of a test that excelled previous standards.

We have testing stations for everything else. Think what the societies for testing materials have done for the progress of the world! Their records are usable forever, in any part of the world, once they are made.

When machines have to be tended, two separate sets of motions must be provided for:

1. The set that the worker uses when he is tending the machine.

2. The set that the worker uses to prepare tools and material for the machine while it does not require his attention.

All machines have to be tended more or less. Even automatic machinery has to have attention, and it is most important here to have motion study, because of the earning value of the machine being lost while it is shut down.

One sees occasionally a machine that can have any and every lever operated without the operator taking a single step, but comparatively few machines are constructed with this in mind.

Machines requiring constant starting and stopping and hand feeding or adjusting should have their various levers so positioned that the "laws of least effort of simultaneous motions" are complied with.

These laws will be discussed under "Variables of the Motion." It is only necessary to say here that motions should be similar on each side of a fore and aft vertical plane passing through the body. It is so necessary to have the motions similar that often counterbalances and springs can be installed to reverse the motion, thus also causing the hardest work to be done in the most convenient direction.

Anything that is used very often can be returned to place better, as well as with less motions, by gravity, or by the application of the gravity by some such means as a string and a weight. It requires some skill to use a wrench, but it requires no skilled motion or thought to return the wrench to its exact resting place with handle pointing in the most economical direction for picking up the next time it is used.

The average machine to-day is designed for a short demonstration of quick output, with less regard for the least percentage of rest required for overcoming fatigue due to continuous operation. With demand will come supply of machines that fulfill all economical motion requirements.

Union Rules

The local rules of some unions are sometimes a hindrance to standardizing motions and thereby increasing output. The higher wages from higher outputs under intensive management soon convert the desirable members, however.

Many unions believe that extremely high outputs per man are against the interests of the union as a whole, on the theory that they may "work all of their members out of a job." Furthermore, they often think that the sacrifice that their one union may make in the world's endeavor to reduce the cost of living generally, is not properly offset by having any one trade or any one locality practicing intensive outputs. A few practical object lessons of the general increase in business resulting from higher wages and simultaneously created lower-production costs will, however, always convince the most prejudiced believer in artificially restricted maximum outputs.

The compensation of workers will not be discussed here, although the basis of compensation does affect motions.

Weight of Unit Moved

Generally speaking, the weight of the unit moved is of three kinds:

1. The weight of that part of the body that is moved.

2. The weight of a tool used, such as a hammer or a trowel.

3. The weight of material used, such as a brick, or the mortar on the trowel.

Other things being equal, the less of the body moved the less fatigue.

The weight that the tool should be is determined by the use of the tool. In the case of a sledge hammer, increased weight means increased efficiency. A twenty-five pound sledge might break a block of granite in halves in five blows, while a ten pound hammer might require one hundred blows. In the case of a trowel, increased weight means decreased efficiency. The heavier the trowel, the greater the fatigue — with no accompanying gain in output.

We have determined that a cutting-out hammer for brickwork should weigh, exclusive of the handle, 3.75 pounds, but that a hammer for drilling plug holes in granite, for making dog holes in heavy stone blocks, should weigh 4 pounds.

The weight of units moved should be standardized.

Example. — There is undoubtedly a certain sized load in a shovel that will enable a first-class man to accomplish the largest output with his maximum effort. Taylor has found his weight to be 21.5 pounds. The size of shovels that should be used should therefore be designated on the instruction card accordingly, and exactly 21.5 pounds should be the standard unit of weight of material shoveled.

Summary

This discussion of the variables of the surroundings, etc., is not detailed — because general discussion is self-evident, and detailed discussion must be too specialized to interest the general reader.

It is only necessary to call attention to the general laws, logical and psychological, which underlie these variables, and their effect on standardizing motions. Each student naturally applies these laws to his own field, and sees for himself the opportunities for further study and application.

CHAPTER IV

VARIABLES OF THE MOTION

A DISCUSSION of variables of the motion opens up a field so large that it is only possible here to attempt to show the method of investigation, and to show that each variable is a necessary factor in making motions standard, leaving to the universities and to properly created and equipped bureaus of the national government the task of reducing motion study to an exact science.

ACCELERATION

In considering acceleration of speed as an element of any motion, we must determine:

1. The amount of acceleration that it is possible or economical to obtain.
2. The means by which the acceleration can be obtained.
3. The effect of the acceleration on
   a. Economy in time required to make the motion.
   b. Economy in time required for rest to overcome the fatigue of having made the motion.

Examples. — 1. Laying brick on a wall from a floor, from the height of the floor level up to three feet eight inches high above the floor, can be done with greatest speed when the brick to be picked up are each maintained at a height of one foot three inches, plus two-thirds the height that the wall is higher than the level of the floor on which the bricklayer stands. The brick to be picked up should never be higher than three feet eight inches under any circumstances.

By maintaining the height of the brick to be laid in this relative position to the height of the wall, the brick will always be in a position that permits the bricklayer to accelerate the speed of transportation of the brick by using the path of the quickest speed.

While bricklayers know nothing about this in theory, they very soon discover it in practice by means of their higher recorded output. Greater outputs will be noticeable as an immediate result of maintaining the brick as nearly as possible at the heights above stated.

2. In laying the filling tiers in any one course, it is most economical to lay the farthest filling tier first and the next farthest tier second, and so on. This enables the bricklayer to accelerate the speed of transportation of the brick up to the instant that it is deposited in the mortar.

The above practice is, of course, much more important on shove-joint work than on brick-and-brick construction.

3. The possible benefits from acceleration should be taken into consideration when determining the sequence in which the tiers shall be laid. The position of the feet of the bricklayer is an important factor in obtaining the acceleration desired. For the best results the feet should

be on separate springy planks, so that the transportation of the brick can be speeded up, in addition to the speed of the arms by simply throwing the body by the aid of the spring of the plank. (See Fig. 13.)

Automaticity

Nearly all often-repeated motions become automatic. This is especially true of motions that require no careful supervision of mind or eye.

The automaticity of motions is of great assistance to the worker whose training and methods conform to standardized motions. This fact makes it necessary to have the apprentice taught the right motions first, last, and always.

The automaticity of motions is a hindrance to the worker who has been accustomed to old-fashioned surroundings, equipment, and tools, and who must adapt himself to standard surroundings.

Example. — A remarkable example of making unnecessary motions as a matter of habit is noticeable in places where the local bricklayers have been accustomed to laying brick that have a decided difference in the top and bottom. This difference makes it necessary to lay no brick upside down on the line. When these bricklayers first worked from packets with the brick in the right position to seize right-side up, they would invariably flop and spin each brick in their hands, first wrong-side up and then back again to the original right-side-up position.

The worker who has been trained wrong also finds it difficult to change his habits when he conforms to standard methods.

Example. — Occasionally we find the bricklayer who will spin or flop a brick that is to be laid in the middle of the wall, although it makes no difference which face of the brick is uppermost in these tiers.

The best way to cure motions that are not necessary but that are made from force of habit is to count the motions aloud, endeavoring to keep down to the standard number of standard motions.

When work is done by both hands simultaneously, it can be done quickest and with least mental effort if the work is done by both hands in a similar manner; that is to say, when one hand makes the same motions to the right as the other does to the left.

Most work is accomplished when both hands start work at the same time, and when the motions can be made at the same relative position on each side of a central fore and aft vertical plane dividing the worker's body symmetrically.

Even if motions cannot be planned to be similar for each hand and performed simultaneously, the plane in which the work is to be done should be carefully located.

If motions are so arranged as to be balanced, as suggested, it is possible not only to take advantage of automaticity, but also to cut down jar to the body. It is on this well-known principle that the shockless jarring machine is built. Balanced motions counteract each other. The result is, less bracing of the body is necessary, and less fatigue ensues.

Combination with Other Motions, and Sequence

A motion may be combined with motions that are (*a*) similar to it, and (*b*) dissimilar to it.

(*a*) If the motions combined are similar to it, advantage must be taken of the automaticity. Care must also be taken that all the motions made in a series of similar motions are necessary. Sometimes one effective motion is preferable to several not so effective.

Examples. — 1. When tapping a brick down to grade with a trowel, one brisk tap will do the work as well as several light taps, and with much less time and effort.

2. If it is necessary to spread mortar on a face tier, one stroke of the trowel will do the work as well as several.

(*b*) If the motions combined are dissimilar, two motions may often be transformed into one.

Example. — The motion used to spread mortar may be combined with the motion used to butter the end of the brick laid just before the mortar was thrown. Thus, the two operations may be transformed into one, and a saving of time and motions will result. In fact, so doing may have other distinct advantages, such as leaving better keying for plastering direct upon the wall.

This subject of combinations of motions can barely be

touched here. Its full treatment involves all other variables, and it can never be considered standardized till each separate motion is a standard.

Cost

The cost of motions, absolute and relative, is a subject too large for any person, firm, or corporation to hope to cover. If complete data are ever to be gathered on it, the cost keeping, recording, and deducing will have to be done by the government.

But all work done by the individual investigator will result in real cost reducing, with increase of output, which is the ultimate purpose of all motion study.

The relative cost of labor and material must be considered.

Examples. — 1. A bricklayer should never stop to pick up dropped mortar. The mortar dropped is not so valuable as the motions necessary to save it.

2. That quality of mortar that is easiest handled by the bricklayer is usually cheapest. The cost of grinding up the lumps in the sand, cement, and lime is less than the cost of the motions necessary to pick the lumps out with a trowel.

3. It is usually cheaper to fill a closer, say less than one-half a brick in size, on the interior tiers, with even the best of cement, than it is to cut a special piece of brick to fit or to walk a few steps to find one the right size. The extra cost of the mortar is negligible compared with the cost of the motions.

The relative cost of motions of higher and lower grades of labor must also be considered.

It is obvious that, other things being equal, it is cheaper to have a low-priced man instead of a high-priced man make the same motion; but only the most careful study can determine all of the motions that could be taken from the high-priced man and allotted to one or more grades of lower-priced men. This can never be wholly or properly accomplished until our present trades, with their inherited conditions and traditions, have been reclassified to meet modern conditions.

In some trades it is very difficult to effect such division of work, as unions are opposed to having anything relating to skilled work done by laborers.

Examples. — 1. In the most highly unionized districts carpenters only are allowed to unload the rough lumber from the cars, and none but carpenters are allowed to transport, lift, and erect, as well as to fabricate it.

2. In bricklaying the case is slightly different. The work of transporting the brick to the place where they are to be laid has always been done by tenders and laborers. The bricklayer never wheels or carries brick. This is a tradition long handed down. Yet he is most jealous that no part of his own work shall be done by a tender or a laborer.

During the time that brick construction was practically without competitors in its field, the bricklayer could insist on his ancient privileges and prosper.

The inroads of concrete, both plain and reinforced, however, have changed conditions, and the bricklayer himself is, more than any other one factor, the cause of many cases of substitutions of concrete for brick.

The architecture of any country is determined by the relative cost of building materials in place, and the history of the world shows that the way to get the most of any one thing used is to make it the lowest in price.

The one thing that will reduce the price of brickwork more than any other is *to reduce the cost of the motions.*

After the laws underlying motion study have all been applied, the cost of motions can still be reduced from one-third to one-half by separating the motions of the bricklayer into at least two classes, such as, for example:

1. Those that require skill.

2. Those that require nothing but strength, endurance, and speed.

Those that require skill should be divided into several classes, according to the amount of skill required; those that chiefly require skill should be handled by mechanics, and those that chiefly require strength, endurance, and speed should be handled by specially trained laborers. This is the only way to enable brickwork to compete with concrete, when all of the architects, engineers, owners, and contractors shall have learned the full possibilities of concrete.

It will be urged that such division of the work of bricklaying will lower the general skill of the bricklayers as a class. Far from it! All operations requiring skill will remain in the hands of the bricklayer, who, escaping all work that unskilled hands could do, will have the more time and energy to devote to the "art" element of his work.

But we are not at this time discussing "brickwork as a lost art" — we cite bricklaying here as an example of the cost of motions, the result of the effects of cost of motions, and of the possibilities and importance of motion study as a method of attack in cost reducing and in standardizing the trades for the greatest possible economy.

What greater service can the bricklayer do both his trade and the people who own or occupy houses than to reduce the cost of the motions in brickwork without reducing his own wages or increasing his hours?

The elimination of wastes is the problem that has been forced to the attention of the entire world to-day, and of America particularly. The elimination of wastes in the trades offers the largest field for savings.

Every trade must be reclassified, and must have the brawn motions separated from the skill motions. Scientific division of the work to be done is as sure to result in higher wages and lower production costs as did F. W. Taylor's separating the planning from the performing.

The reason that our country is not astounded and confused at the appalling unnecessary loss to its inhabitants on account of unnecessary, wasteful, and improper motions of its workers is due to ignorance of the existence of this loss, and to ignorance of any method of eliminating it.

The loss due to the present classification of the trades alone is probably more than sufficient to pension, under full pay, one-half of the workers of the country; is certainly enough to enable all of the women and children in the trades to remain out of the trades and be paid at their regular wages.

While such action is not even recommended, the illustration is used to emphasize the enormous waste going on daily and yearly.

That we go on year after year submitting to this waste because our present trades are handled in accordance with ancient conditions entirely out of place in our present civilization, is no longer necessary and without excuse.

Let the government call its scientific managerial experts together and make a test of one trade, reclassify it, and publish the data. The object lesson thus presented will cause to be taken the necessary further steps to remedy the present system of handling the trades. The workers will each be able to earn higher wages when the unions see that they are benefited, and the labor interests will coöperate. The cost of living will be reduced as by no other means, and all this by scientifically reclassifying the trades!

Direction

In most cases, the direction of a motion that is most economical is the one that utilizes gravitation the most.

Oftentimes delivering material to a high-priced workman by leaving the material in a high position also makes easy unloading for the low-priced workman.

Example. — Stacking up packs 2 feet high saves motions, and saves stooping when the laborer unloads his trucket. (See Fig. 21.)

"Direction" admirably serves as an illustration of the

Fig. 21.—Trucket for storing reserve packs stacked up 2 ft. high, to save stooping when it is being unloaded.

close interrelation of the variables. It is closely connected with "path." It involves discussions of anatomy, acceleration, and speed. It demands consideration of all variables of surroundings, equipment, and tools.

The best "direction of motion" is not only important in itself for increase of output; it must also be kept constantly in mind in standardizing the placing of both materials and men.

Effectiveness

Effectiveness has been touched upon in discussing "combination with other motions."

An effective motion is one that produces the desired result. Oftentimes whole processes, methods, and operations can be so changed as to make the succeeding motions much more effective.

Example. — The introduction of the fountain trowel,

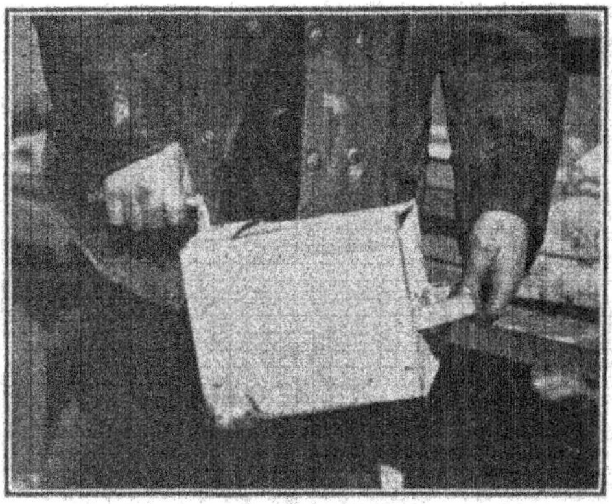

Fig. 22. — The Fountain Trowel.

used in connection with an ordinary trowel, made each motion in handling mortar much more effective. (See Figs. 19, 22.)

Foot-Pounds of Work Accomplished

After all, a human being or a work animal is a power plant, and is subject to nearly all the laws that govern and limit the power plant. It is a law of motion study that, other things being equal, the less number of foot-pounds of work done by the workman, the smaller percentage of working hours he must devote to rest to overcome fatigue.

It is therefore of great importance in obtaining the largest possible output that the work shall be so arranged and the workman so placed that he can do his work with the least possible amount of foot-pounds of work done per unit of output accomplished. This is where the philanthropic employer has often been rewarded without knowing it. In his desire to make conditions such that the workman was most comfortable while working, he reduced the number of foot-pounds of work to that which was absolutely necessary to do the work. He surrounded the workman with conditions that enabled him to have no fatigue, except that which was acquired from the motions of the work itself. He made conditions such that the workman was enabled to overcome the fatigue from his motions in the quickest possible time. (See Fig. 23.)

Inertia and Momentum Overcome

There are two ways by which the amount of inertia and momentum may be reduced.

1. By standardizing surroundings and equipment so that the inertia and the momentum are limited to practically that of the materials, and not the materials plus arms and body.

Example. — Picking up ninety pounds of brick at one lifting.

2. By so standardizing motions that as few starts and stops as possible occur from the time the material leaves the stock pile till the time it is in its final resting place in the work.

Example. — In laying brick by the "pick-and-dip" method on face tiers, a brick is lifted in one hand and a trowel full of mortar in the other. The brick must come to a full stop in the bricklayer's hand while the mortar is being laid and the bed prepared, and then move to its final resting place, unless brick and mortar are dropped in two different places.

In laying brick by the "stringing-mortar" method, the mortar is laid and the bed prepared before the bricks are lifted. The brick are conveyed from the pack to the wall without interruption or delay.

Standard methods of performing work may enable the worker to utilize the momentum.

Example. — If the bricks are conveyed from the stock platform or pack to the wall *with no stops*, the momentum can be made to do valuable work by assisting to shove the joints full of mortar. If, instead of being utilized, the momentum must be overcome by the muscles of the bricklayer fatigue, not full joints, will result.

The ideal case is to move the brick in a straight path and make the contact with the wall overcome the momentum.

Length

A general rule of motion economy is to make the shortest motions possible.

Eliminating unnecessary distances that workers' hands

Fig. 23.—Common type of Trestle Horse Staging. Bricklayers cannot be expected to lay as many brick per day when working in this position as when standing up between a wall and a stock platform, each 2 ft. high.

and arms must travel, will eliminate miles of motions per man in a working day as compared with usual practice.

Example.—Put the wheelbarrow body as close as possible to the pile that is to be put into it, so that the distance the packets are carried from the pile to the barrow, or the sand from the pile to the barrow, will be the shortest distance possible.

Of the necessary distance to be walked or reached, have as much of it as possible done by the low-priced man, and have as little of it as possible done by the high-priced man.

Example. — With brick, have the tender put the pack of brick as near the final resting place of the brick as conditions will permit, so that when the high-priced man picks up a pack of, say, eighteen bricks, he requires a short motion only.

Have the high-priced worker always use first the stock that is nearest, this rule requiring the shortest motions in conveying the stock to its final resting place.

Example. — In picking up brick from a packet or a scaffold the nearest brick should be picked up first. The brick that are farthest away serve as a reserve stock pile, to be picked up only in the emergency of not having any others nearer to pick up. It may be that the brick farthest away may not need to be used on that piece of work at all, or at least their place will not be occupied so many times by bricks to be transported with longer motions.

Standard tools, equipment, and surroundings are essential if length of motions is to be made standard.

As already said when discussing clothes, the workman of the present should have even his overalls, belt, and clothes so designed that they will hold the different kinds of tools that are oftenest used, so that they may be picked up in the shortest time — that is, with pockets for nails, clips, clamps, etc. The tools should be so placed that the least and shortest motions can be used after they are picked up, as cartridges are placed in a cartridge belt.

Necessity

The necessity of the motion is such an important variable that an investigator is tempted at first glance to divide all motions into necessary and unnecessary, and to eliminate with one stroke those that appear to him unnecessary. A more thorough investigation will be apt to prove that no such summary elimination is advisable.

A motion may be an unnecessary motion in a necessary sequence, or it may be a necessary motion in a certain sequence, but the whole sequence may be unnecessary or inadvisable.

Example. — In opening a paper bag of cement the average untrained laborer usually cuts the bag in two and removes the paper in several pieces and with many motions. The correct way is to cut the bottom with a shovel and pull the bag upward in one piece by grasping the bag just above the string.

This example shows both how motions may be unnecessary in themselves and how they may belong to a sequence that is unnecessary.

The only final solution as to the necessity of a motion will come when the trades are completely standardized.

It is impossible to determine whether or not a motion is absolutely necessary until the method of doing the work in which it is used is standard.

Examples. — 1. Motions which were relatively proved necessary in laying brick by the "pick-and-dip" method or "stringing-mortar" method, the brick being lifted from the stock platform, became absolutely unnecessary when the "packet-on-the-wall" method of handling brick was adopted.

2. The same thing is true of motions eliminated by handling mortar in a fountain trowel.

The final solution of the problem of necessity of motions will be discussed later, though the subject is so large that no amount of discussion could do more than touch it.

Path

The determination of the path which will result in the greatest economy of motion and the greatest increase of output is a subject for the closest investigation and the most scientific determination. Not until data are accumulated by trained observers can standard paths be adopted. The laws underlying physics, physiology, and psychology must be considered and followed. In the meantime, merely applying the results of observation will reduce motions and costs and increase output to an amazing degree.

The path most desirable is usually that which permits gravitation to assist in carrying the material to place.

Example. — We have found that the most economical height for laying brick is twenty-four inches above where the bricklayer stands, while it is most economical to pick the brick from a height about three feet above where the bricklayer stands; that is, about one foot higher than the top of the wall where the brick is to be laid.

The path is affected by the direction that the material is to be shoved as it moves into its final resting place.

Examples. — When the packet is placed on the wall it should be placed so that the brick can be picked up and moved in a comparatively straight line with the direction that the brick will be shoved for filling a joint.

In theory the ideal path would be in a line of quickest speed from the stock platform to the wall.

In practice it is seldom that the most economical path for carrying a brick or mortar from the stock platform to the wall is exactly a straight line from one to the other. It will generally be most economical to move the brick in the path that will bend the arms the least and that will permit almost a swing from the shoulder.

Playing for Position

Each motion should be made so as to be most economically combined with the next motion, like the billiard player who plays for position.

The direction in which a motion is made may affect the time required for a subsequent motion.

Example. — In laying brick the motion of placing the

mortar for the end joint can be done quickest if it is done in the direction of the next motion, such, for example, as the next motion that puts the trowel in the position to cut off the hanging mortar.

The sequence of motions in bricklaying, that determines when the particular motion is to be made that puts the mortar in the end joint, depends upon whether the "pick-and-dip" or the "stringing-mortar" method is used.

When the motions are made in the correct sequence, many of them can be combined so that two, and in some cases three, motions can be made as one motion, in but little more time than is required for one motion.

Example. — Cutting off mortar, buttering the end of the laid brick, and reaching for more mortar all as one motion, in the "pick-and-dip" method.

Speed

Usually, the faster the motions, the more output. There are other advantages to speed of motions besides the fact that they require less time. Speed increases momentum, and this momentum may be utilized to do work.

Example. — The momentum of the brick helps to shove the mortar better into the joint.

Again, high outputs are generally the result of the habit of speed in motions. Habits of speed are hard to form, and they are hard to break.

Next to fewest motions, speed of motions is the most important factor of high record of outputs.

The list of variables here given makes no claim to being complete. The field of study is so immense that it is impossible as yet to give a complete and detailed method of attack.

It will be noted in reading the discussion of the variables that it has been found extremely difficult to handle each one separately. It is needless to tell the student, the investigator, the cost-reducing manager, that, difficult as the task is, for the best results each variable must be studied alone. The effects of all variables but one must be eliminated, or, better perhaps, all variables but one must be maintained constant.

Quicker results may often be obtained by studying several variables simultaneously, and for short jobs this may be advisable. But for long jobs of repetitive work there is no way so accurate and satisfactory as studying one variable at a time.

CHAPTER V

PAST, PRESENT, AND FUTURE OF MOTION STUDY

WORK ACCOMPLISHED

CONSIDERED in relation to the time during which it has been applied to the trades, scientific motion study can show most satisfactory results.

The workers in the field as well as in the office have been quick to appreciate and adopt the new methods suggested by motion economy.

This has been especially the case in the crafts. Nearly every proficient workman loves his trade. He loves the joy of achievement. He can achieve most when useless motions have been eliminated for him, and he welcomes improvements, as the bricklayers have welcomed the brick coming right side up on the packet.

MAGNITUDE OF WORK TO BE DONE

To the casual reader it may seem that the task of evolving standard practice from usual present practice, and from the best practice, is simply a case of observing, recording, and eliminating. The student will see that it requires the closest concentration to do even the necessary scientific observing and recording, while to deduce and systematize standard motions for any one trade would furnish a life work for several trained scientists.

It is a difficult task for an inexperienced or untrained observer to divide an operation correctly into its motions. Enumerating the variables that affect each motion is a task big enough to satisfy the most ambitious student of waste elimination.

Value of Charts

We have found it helpful in recording our observations to use charts. Some such form as that shown on pages 88 and 89 is used.

This chart is one made during an observation of bricklaying before the invention of the packet, the packet scaffold, and the fountain trowel.

The operation of laying a brick was divided into the motions of which it consisted (column 1). The usual (present) practice of the time (given as "the wrong way," column 2) showed the units into which the operation was divided. The best practice of the time ("the right way," column 3, now obsolete) was charted in such a way that its relation *from a motion standpoint* to the usual practice was clearly shown.

Column 4 shows how the usual practice may be transformed into the best practice. It would serve as an instruction card to the workman, showing him not only where his method needed to be improved but also exactly how to improve it.

SAMPLE CHART USED ON PICK-AND-DIP METHOD, EXTERIOR BRICKWORK, BEFORE THE DAYS OF THE PACKET

Operation No.	The Wrong Way, Motions per Brick, ¼ ½ ¾ 4/4	The Right Way, Motions per Brick, ¼ ½ ¾ 4/4	Pick-and-Dip Method. The Exterior 4 inches (Laying to the Line).
1	Step for mortar	Omit	On the scaffold the inside edge of mortar box should be plumb with inside edge of stock platform. On floor the inside edge of mortar box should be 21 in. from wall. Mortar boxes never over 4 ft. apart.
2	Reaching for mortar	4/4	Do not bend any more than absolutely necessary to reach mortar with a straight arm.
3	Working up mortar	Omit	Provide mortar of right consistency. Examine sand screen and keep in repair so that no pebbles can get through. Keep tender on scaffold to temper up and keep mortar worked up right.
4	Step for brick	Omit	If tubs are kept 4 ft. apart, no stepping will be necessary on scaffold. On floor keep brick in a pile not nearer than 1 ft. nor more than 4 ft. 6 in. from the wall.
5	Reach for brick	Included in 2	Brick must be reached for at the same time that mortar is reached for and picked up at exactly the same time the mortar is picked up. If it is not picked up at the same time, allowance must be made for operation.
6	Pick up right brick	Omit	Train the leaders of the tenders to vary the kind of brick used as much as possible to suit the conditions; that is, to bring the best brick when the men are working on the line.
7	Mortar box to wall	4/4	Carry stock from the staging to the wall in the straightest possible line and with an even speed, without pause or hitch. It is important to move the stock with an even speed and not by quick jerks.
8	Brick pile to wall	Included in 7	Brick must be carried from pile to wall at exactly the same time the mortar is carried to the wall, without pause or jerk.

9	Deposit mortar on wall		Included in 7	If pause is made, this space must be filled out. If no pause is made, it is included in No. 7.
10	Spreading mortar		Omit	The mortar must be thrown so as to require no additional spreading and so that the mortar runs up on the end of the previous brick laid, or else the next two spaces must be filled out.
11	Cutting off mortar		Omit	If the mortar is thrown from the trowel properly, no spreading and no cutting is necessary.
12	Disposing of mortar		Omit	If mortar is not cut off, this space is not filled out. If mortar is cut off, keep it on trowel and carry back on trowel to box, or else butter on end of brick. Do not throw it on mortar box.
13	Laying brick on mortar		¼	Fill out this space if the brick is held still while mortar is thrown on wall. When the brick is laid on mortar it presses mortar out of joints; cut this off only at every second brick. It takes no longer to cut mortar off two bricks than one.
14	Cutting off mortar	Every 2d brick	½	
15	Disposing of mortar	Butter end joint	¼	When this mortar is cut off it can be used to butter that end of the last previous brick laid, or it can be carried on the trowel back to the box.
16	Tapping down brick		Omit	If the mortar is of the right consistency, with no lumps in it, and the right amount is used, and the bricks are as wet as possible without having them run, no tapping with the trowel will be necessary.
17	Cutting off mortar		Omit	If the brick must be tapped, hit it once hard enough to hammer it down where it belongs. Do not hit the brick several light taps where one hard tap will do.
18	Disposing of mortar		Omit	Do not cut off the mortar oftener than every second brick, and, when you do cut it off, do not let it fall to the ground; save it; keep it on the trowel, and do not make another motion by throwing it at the box. Carrying it to the box does not count as another motion.
	18		4½	Total number of motions per brick.

This chart, together with a plan showing the workman where he should put the stock and where he should place his feet (Fig. 14), and with pictures showing how he should lay the brick, etc., proved most successful for instruction as well as for recording.

At first glance this chart, and the others like it, which we used at that time, seem very crude. In fact, compared to what has since been done to standardize operations, they *are* crude. But they mark a distinct phase of motion study. They show plainly, as careful reading will prove, that an earnest study of motions will automatically promote the growth of the study.

For example, study of column 4 in the sample chart given led to the invention of the packet scaffold, the packet, the fountain trowel, and several other of the best devices, and the "packet-on-the-wall" method now used in brickwork.

These inventions in their turn necessitated an entirely new set of motions to perform the operation of laying a brick.

So, likewise, the progression also went on before the days of conscious motion study: observation, explanation, invention, elimination, and again observation, in an upward helix of progress.

The great point to be observed is this: Once the variables of motions are determined, and the laws of underlying motions and their efficiency deduced, conformity to these laws will result in standard motions, standard tools, standard conditions, and standard methods of performing the operations of the trades.

Conformity to these laws allows standard practice to be attained and used. If the standard methods are deduced before the equipment, tools, surroundings, etc., are standardized, the invention of these standard means is as sure as the appearance of a celestial body at the time and place where mathematics predicts that it will appear.

It is as well to recognize first as last that real progress from the best present method to the standard method can never be made solely by elimination. The sooner this is recognized the better. Elimination is often an admirable makeshift. But the only real progress comes through a reconstruction of the operation, building it up of standardized units, or elements.

It is also well to recognize the absolute necessity of the trained scientific investigator. The worker cannot, by himself, arrange to do his work in the most economical manner in accordance with the laws of motion study. Oftentimes, in fact nearly always, the worker will believe that the new method takes longer than the old method. At least he will be positive that many parts, or elements, of the process when done under the new method take longer than under the old style, and will not be in sympathy with the scheme because he is sure that the new way is not so efficient as his old way. All of which shows that the worker himself cannot tell which are the most advantageous motions. He must judge by the fatigue that he feels, or else by the quantity of output accomplished in a given time. To judge by the quantity of output accomplished in a given time is more of a test of effort than a test of motion study, and oftentimes that element that will produce the most output is the one that will cause the least fatigue.

The difference in amount of merit between any two methods can perhaps be best determined by timing the elements of the motions used in each. This is the method of attack usually accepted as best, because it separates each motion into its variables and analyzes them one at a time. It is out of the question to expect a workman to do such timing and to do his work at the same time. Furthermore, it is an art in itself to take time-study observations, an art that probably takes longer to master than does shorthand, typewriting, telegraphy, or drafting.

Few workers have had an opportunity to learn the art of making and using time-study observations, because our school educators have not had any mental grasp of the subject themselves. Add to the difficulties to be overcome in acquiring the knowledge of observing, recording, and analyzing the time-study records, the knowledge necessary to build up synthetically the correct method with each element strictly in accordance with the laws of motion economy each by itself and when used together in the particular determined sequence, and you will see the reason why the worker by himself has not devised, cannot, and never will be expected to devise, the ultimate method of output. It does not then, after all, seem so queer that the workman's output can always be doubled and oftentimes more than tripled by scientific motion study. Again, scientifically attained methods only can become Ultimate methods.

Any method which seems after careful study to have attained perfection, using absolutely the least number of most effective, shortest motions, may be thrown aside when a new way of transporting or placing material or men is introduced. It is pitiful to think of the time, money, strength, and brains that have been wasted on devising and using wonderfully clever but not fundamentally derived methods of doing work, which must inevitably be discarded for the latter.

The standardizing of the trades will utilize every atom of such heretofore wasted energy.

The standardizing of the trades affords a definite best method of doing each element.

Having but one standard method of doing each element divides the amount of time-study data necessary to take by a number equal to the number of different equally good methods that could be used.

The greatest step forward can be made only when time-study data can be made by one and used by all. A system of interchange and coöperation in the use of the data of scientific management can then be used by all persons interested.

This reduction and simplification of taking time study is the real reason for insistence upon making and maintaining standards for the largest down to the smallest insignificant tool or device used.

Much toward standardizing the trades has already been done. In this, as in almost countless other lines of activity, the investigator turns oftenest with admiration to the work of Frederick W. Taylor. It is the never-ceasing marvel concerning this man that age cannot wither nor custom stale his work. After many a weary day's study the investigator awakes from a dream of greatness to find that he has only worked out a new proof for a problem that Taylor has already solved.

Time study, the instruction card, functional foremanship, the differential rate piece method of compensation, and numerous other scientifically derived methods of decreasing costs and increasing output and wages—these are by no means his only contributions toward standardizing the trades whose value it would be difficult to overestimate; they are but a few of the means toward attaining standards which have been placed by Taylor, their discoverer, within the hands of any man willing to use them.

Future Work in Standardizing the Trades

The great need to-day in standardizing the trades is for coöperation. In other times all excellent methods or means were held as "trade secrets," sometimes lost to the world for generations until rediscovered. The day for this is past. Thinkers of to-day recognize that the work to be done is so great that, given all that every one has accomplished and is accomplishing, there is room and to spare for every worker who cares to enter the field. Co-operation and team work is the crying need.

Conservation and comparison of knowledge, experiments, data and conclusions are what we need. The various engineering journals are to be commended for recognizing the importance of this, and for furnishing an excellent means for recording and spreading much needed information.

The ideal conservator of knowledge in this, as in all other branches, would be the United States government. The government should maintain a permanent bureau, with experiment stations, as is done with the Department of Agriculture.

Individual investigators, corporations, and colleges, all would be willing to turn over the results of their work to such a government bureau. The colleges would coöperate with such a bureau, as do the agricultural colleges with the Department of Agriculture. The bulletins of such a bureau would be invaluable to the men in the trades, as are the agricultural bulletins to the farmers.

The Department of Agriculture is an excellent model. The form for a department or bureau of trades is all at hand. It is only necessary to translate the language of agriculture into the language of labor. It is only through such a bureau that the trades can formally be standardized.

Such a bureau would have two main tasks: (1) To sub-classify the trades; (2) To standardize the trades.

The first task should be successfully completed before the second is undertaken.

We have spoken briefly, in considering cost of motions, o the necessity of separating those motions that require skill from those that require nothing but strength and endurance.

This sub-classifying of the trades according to the types or grades of motions that they use, or according to the brawn, brain, training, and skill required to make the motions, will cut down production costs. It will raise the standards of all classes. It will do away with differences between employers and employees. It will eliminate unnecessary waste. It will raise the wages of all workers. It will reduce the cost of living.

We might call such a sub-classification as desired a "functional" classification of the trades.

For example, for brickwork we recommend five classes:

Class A. — Ornamental and exterior face brick and molded terra cotta.

Class B. — Interior face tiers that do not show at completion, where strong, plumb, and straight work only is needed.

Class C. — Filling tiers where only strength is needed.

Class D. — Putting fountain trowels and brick packs on the wall near the place, and in the manner where the other three classes can reach them with greatest economy of motion.

Class E. — Pack loaders, brick cullers, and stage builders.

The pay of the A and B classes should be considerably higher than is customary for bricklayers. The pay of the C, D, and E classes should be lower than is customary for bricklayers, but much higher than the pay of laborers. This classification will raise the pay of all five classes higher than they could ever obtain in the classes that they would ordinarily work in under the present system, yet the resulting cost of the labor on brickwork would be much less, and each class would be raised in its standing and educated for better work and higher wages.

In the case of brickwork this new classification is a crying necessity, as the cost of brickwork must be reduced to a point where it can compete with concrete. Improvements in making, methods of mixing, transporting, and densifying concrete in the metal molds of to-day have put the entire brickwork proposition where it can be used for looks only, because for strength, imperviousness, quickness of construction, lack of union labor troubles, and low cost, brickwork cannot compete with concrete under present conditions.

Having sub-classified the trades, the second step is to standardize them.

And both classification and standardization demand motion study.

The United States government has already spent millions and used many of the best of minds on the subject of motion study as applied to war; the motions of the sword, gun, and bayonet drill are wonderfully perfect from the standpoint of the requirements of their use. This same study should be applied to the arts of peace.

It is obvious that this work must and will be done in time. But there is inestimable loss in every hour of delay. The waste of energy of the workers in the industries to-day is pitiful. But it is far more important that the coming generation of workers should be scientifically trained.

The science of management of the future will demand that the trades be taught in accordance with the motion standards of a United States Bureau of Standardization of Mechanical Trades. The present method of teaching an apprentice is the most unbusinesslike event that takes place in any of our industrial institutions.

We have never heard of a trades school, manual training school, or technical school that makes any attempt to solve questions of motion study. The usual process is to teach a student or apprentice to do his work well first, and after he has finally accomplished the art of making or doing the thing in question, then to expect him to learn to do it quickly. This process is a relic of the dark ages. A novice should be taught to do what he is trying to do with certain definite motions, and to repeat the operation until he is able automatically to use the standard motions and do good work.

If an apprentice bricklayer, blacksmith, or tool sharpener, for example, is not instructed to count his motions when doing a certain piece of work, he will surely get into the habit of making extra motions that cannot be omitted later without almost as much effort as that spent in learning the trade. There is little incentive for an old mechanic to teach a boy so that he will excel his teacher, and perhaps run him out of a job about the time that he, the apprentice, becomes expert.

One of the most common causes for neglecting the important subject of motion study is that the boss of the establishment is not himself really a master of the trade that is being taught, or, if he was master once, has forgotten it because there are no books or systems that have so described, charted, and illustrated his trade as to refresh his memory.

Again the teacher is often a mechanic who is not trained to impart what knowledge he has, has never studied pedagogy, and is expected to do a full day's work at the same time that he is teaching his apprentice.

The arts and trades of human beings should be studied, charted, photographed, and motion-pictured, and every employer, apprentice, and student should be able to receive bulletins of his trade for a sum equal to the cost to a farmer of a bulletin from the Department of Agriculture instructing how to increase the outputs of cows, hens, and bees.

One great aid toward cutting down the work of every one out of the trades as well as in, would be the standardizing of our written alphabet to conform to the laws of motion study. The most offhand analysis of our written alphabet shows that it is full of absolutely useless strokes, all of which require what are really wasted motions.

Consider the single example of the first stroke on the first letter of each word. Here is a motion that can be eliminated wholly. While its existence is necessary in type that represents handwriting or imitates engraved plate work, and in enameled separate letters of window signs, its adoption and use in handwriting is of no purpose and is wrong from the standpoint of motion economy.

Each letter of our written alphabet is a natural deviation from our printed alphabet that is the result of leaving the pencil on the paper.

Now the time has arrived for revising our written language by means of a new scientifically invented alphabet specially devised for the purpose of securing clearer writing, made of connected letters, each designed of itself and in connection with all the other letters, so that it conforms to the laws of motion economy. This is not a suggestion that we should adopt stenographic signs for words or sounds, although a general knowledge of one standard stenographic system would also be a great benefit to a nation.

The suggestion is, that in as much as it is the aim of our nation that all citizens should be able to read and write, a new written alphabet should be devised for us that shall conform to the laws of motion study, — that we all can increase either our outputs in writing or else that we all may be able to do such writing as we are obliged to do in less time.

It is to be hoped that an international society of highly trained educators, similar to those composing the Simplified

Spelling Board, may be called together, as was the Simplified Spelling Board, to give this matter immediate attention. A written alphabet for all languages of the world should be determined and used not only by the users of each language, but also by the societies advocating and promulgating such world's second or international languages as Volapük and Esperanto.

One great drawback to the more rapid progress of any artificial or second language has been the difficulty of reading the correspondence between enthusiasts who were proficient in speaking their thoroughly agreed upon international language.

It would not be desirable to abandon our present written alphabet. There are now literally hundreds of different styles of lettering that all can read, yet how few of them can any of us make with pen or pencil.

To add one more style of lettering to the now existing hundreds could scarcely be considered as confusing by even those who are constitutionally opposed to changes in anything.

Therefore, there should be devised one more style of lettering, specially adapted to cutting down the time of writing and adding to the general legibility when written quickly.

Let this be our second written language. Let us use the present system and the new one. Let the generations to come have the benefit of the application of science to their future writing, and let the present style be also used, provided it does not die the natural death in the combat of the survival of the fittest.

We may have to wait for international coinage, international postage stamps, international courts, international arbitration, and international weights and measures; but there can be no reason for not having an international system of written alphabetical characters, and while having it let us decide in favor of that system that fulfills the requirements of motion study, both of the hand in making, and of the eye in reading.

THE FIRST STEPS

In the meantime, while we are waiting for the politicians and educators to realize the importance of this subject and to create the bureaus and societies to undertake and complete the work, we need not be idle. There is work in abundance to be done.

Motion study must be applied to all the industries. Our trade schools and colleges can:

1. Observe the best work of the best workers.
2. Photograph the methods used.
3. Record the methods used.
4. Record outputs.
5. Record costs.
6. Deduce laws.
7. Establish laboratories "for trying out laws."
8. Embody laws in instructions.
9. Publish bulletins.
10. Coöperate to spread results and to train the rising generation.

This is the era — *now*. We have a scientific method of attack, and we have also scientific methods of teaching.

The stereoscopic camera and stereoscope, the motion picture machines, and the stereopticon enable us to observe, record, and teach as one never could in the past.

The following motion study pictures, charts, and diagrams are typical and have been used for teaching journeymen and apprentice bricklayers our standard methods.

PICK-AND-DIP METHOD — WORKING RIGHT TO LEFT

. 24. — Spreading mortar on exterior face tier.

Fig. 25. — Cutting off mortar before the brick is laid on exterior face tier.

. 26. — Buttering the end of the laid brick on the exterior face tier.

Fig. 27. — Cutting off the mortar after the brick is laid on the exterior face tier.

PICK-AND-DIP METHOD — WORKING RIGHT TO LEFT (Continued)

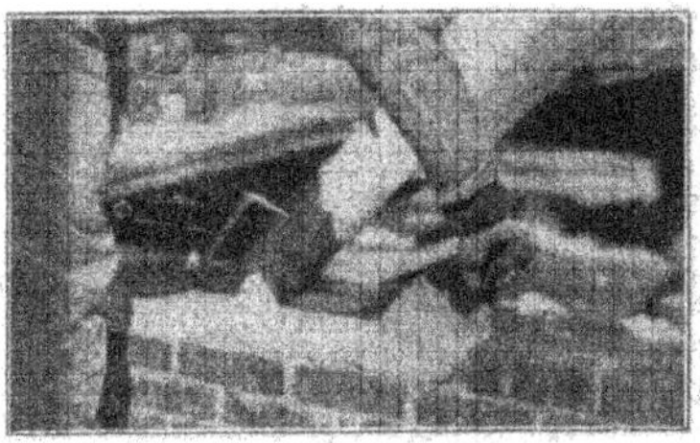

Fig. 28. — Throwing mortar on interior face tier.

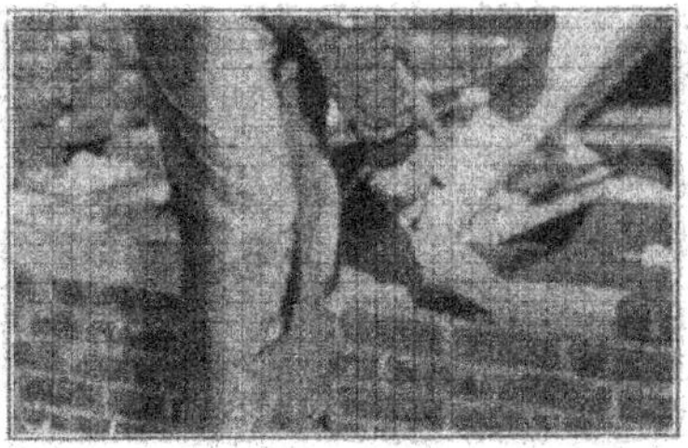

Fig. 29. — Spreading mortar on interior face tier.

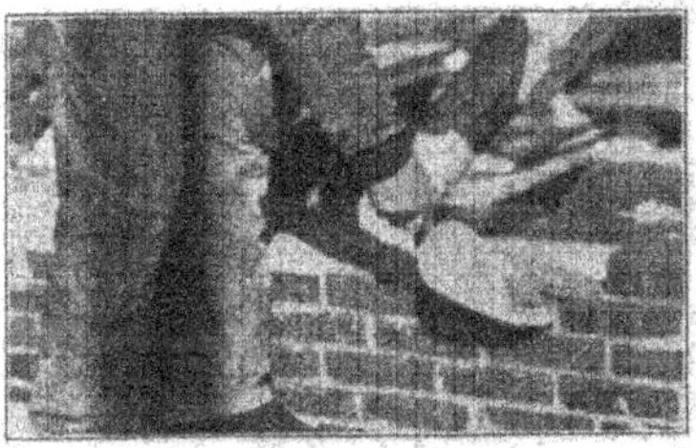

Fig. 30. — Cutting off mortar before brick is laid.

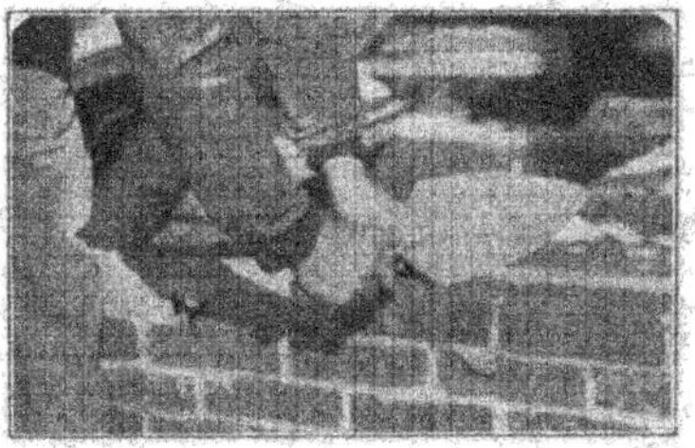

Fig. 31. — Tapping down brick after laying on interior face tier.

Fig. 32. — Cutting off mortar after brick is laid on interior face tier (here working left to right).

Fig. 33. — Handling mortar for two brick at one time on interior face tier.

STRINGING-MORTAR METHOD

Fig. 34.—Working right to left.—Spreading mortar on exterior face tier.

Fig. 35.—Working left to right.—Buttering end of laid brick on exterior face tier.

Fig. 36.—Working left to right.—Cutting off mortar after brick is laid on exterior face tier.

Fig. 37.—Working right to left.—Spreading mortar on the interior face tier.

Fig. 38.—Working left to right.—Cutting off mortar after brick is laid on exterior face tier.

Fig. 39.—Working left to right.—Spreading mortar on the interior face tier.

STRINGING-MORTAR METHOD (Continued)

Fig. 40.—Working left to right.—Spreading mortar on the interior face tier.

Fig. 41.—Working left to right.—Throwing mortar on the interior face tier.

Fig. 42.—Working right to left.—Buttering right-hand end of brick in hand to be laid on interior face tier.

The "pack-on-the-wall" method is the latest development and is an actual direct result of motion study. It has again changed the entire method of laying brick by reducing the kind, number, sequence and length of motions. It reduces the fatigue of the bricklayer and he is therefore able to make more rapid motions.

SAMPLE CHART USED FOR STRINGING-MORTAR METHOD ON THE FILLING TIERS, BEFORE THE DAYS OF THE PACK-ON-THE-WALL METHOD

Operation No.	The Wrong Way, Motions per Brick, ¼ ½ ¾ 4/4	The Right Way, Motions per Brick, ¼ ½ ¾ 4/4	Stringing-Mortar Method. The Center of the Wall.
1	Step for mortar	Omit	
2	Reaching for mortar	¼	Mortar enough for four brick at a time = ¼ motion per brick.
3	Working up mortar	Omit	
4	Step for brick	Omit	
5	Reach for brick	½	One brick in each hand = ½ motion per brick.
6	Pick up right brick	Omit	
7	Mortar box to wall	¼	Mortar for four bricks at once = ¼ motion per brick.
8	Brick pile to wall	½	One brick in each hand = ½ motion per brick.
9	Deposit mortar on wall	Included in 7	
10	Spreading mortar	¼	Spreading mortar for four brick = ¼ motion per brick.
11	Cutting off mortar	Omit	
12	Disposing of mortar	Omit	
13	Laying brick on mortar	Included in 8	
14	Cutting off mortar	Omit	If the brick is not shoved too far on the bed of mortar, the mortar will be shoved just to the top edge of the brick, and no cutting of mortar above top of brick will be necessary.
15	Disposing of mortar	Omit	
16	Tapping down brick	Omit	
17	Cutting off mortar	Omit	
18	Disposing of mortar	Omit	
	18	1¾	Total number of motions per brick.

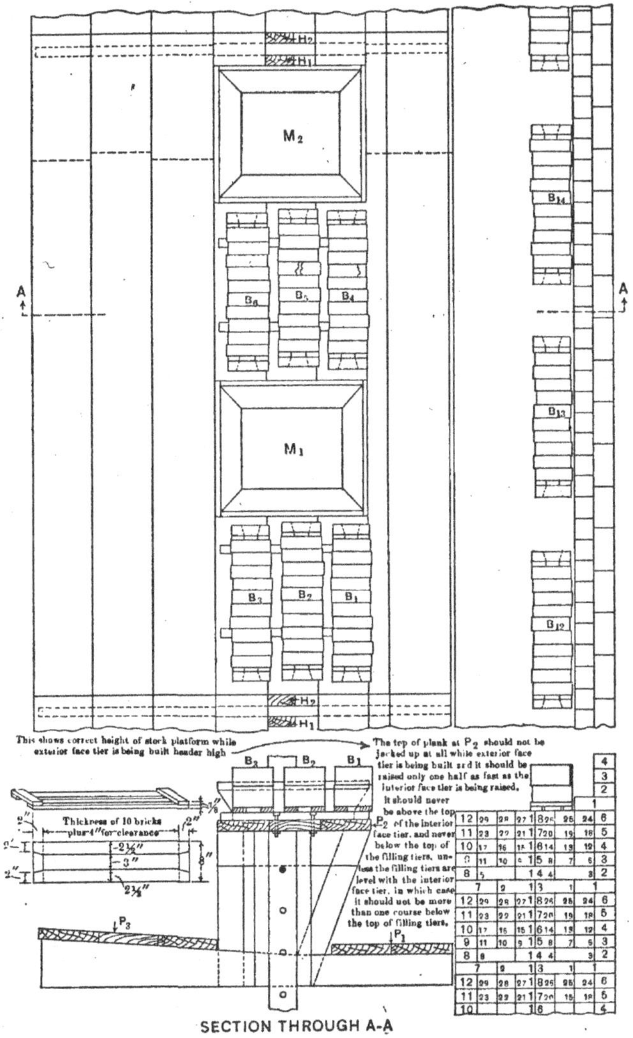

Fig. 43.—Location of the pack on the wall while building the exterior face tiers. Pack-on-the-wall method.

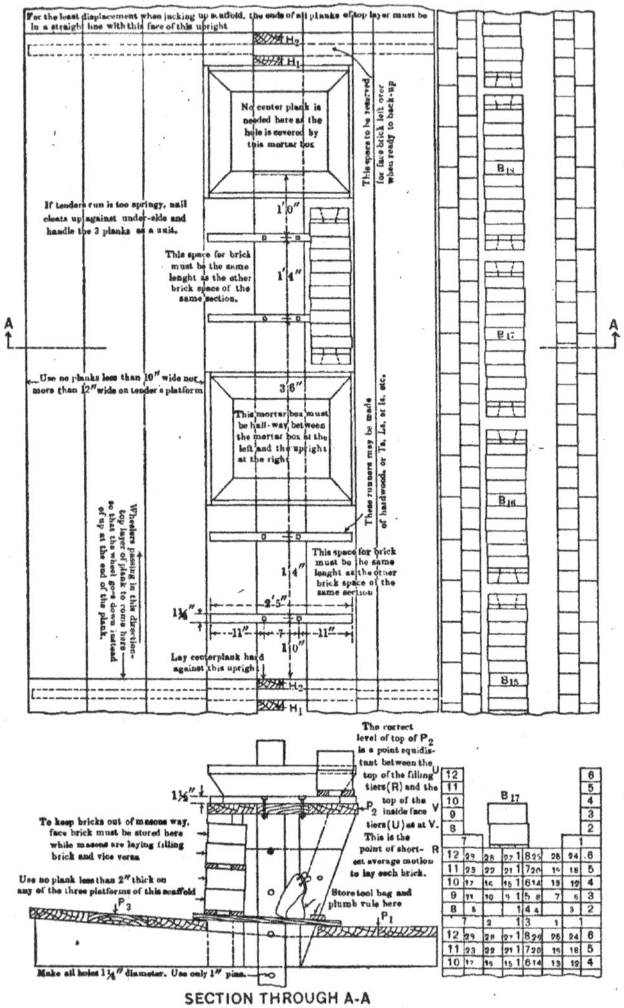

Fig. 44.—Location of the packs while building the interior face and filling tiers. Pack-on-the-wall method.

The economic value of motion study has been proved by the fact that by means of it workmen's outputs have been more than tripled, production costs lowered, and wages increased simultaneously.

This book is written for the express purpose of calling to the attention of the nation that what has been done in a few trades can be done in each and every trade.

The most important matter before the public to-day is the creation and operation of a department at Washington for discovering, collecting, conserving and disseminating data relating to Taylor's method of Intensive Management commonly called Scientific Management.

INDEX

Spotted something that isn't quite right?

Let us know and we'll fix it for future reprints.
And as a thank you for your diligence, we'll
send you a free Sothis book of your choice.

publisher@sothis.press

This page intentionally left blank

Except of course for that line above. And this one.
And whilst we have your attention, perhaps you'd
like to follow us on Instagram @sothis.press. Go on.

Made in United States
Orlando, FL
04 November 2026

90207775R00083